THE NEED TO KNOW: INFORMATION SHARING LESSONS FOR DISASTER RESPONSE

HEARING

BEFORE THE

COMMITTEE ON GOVERNMENT REFORM

HOUSE OF REPRESENTATIVES

ONE HUNDRED NINTH CONGRESS

SECOND SESSION

MARCH 30, 2006

Serial No. 109–143

Printed for the use of the Committee on Government Reform

Available via the World Wide Web: http://www.gpoaccess.gov/congress/index.html
http://www.house.gov/reform

U.S. GOVERNMENT PRINTING OFFICE

27–721 PDF   WASHINGTON : 2006

CONTENTS

THE NEED TO KNOW: INFORMATION SHARING LESSONS FOR DISASTER RESPONSE

THURSDAY, MARCH 30, 2006

House of Representatives,
Committee on Government Reform,
Washington, DC.

The committee met, pursuant to notice, at 10:13 a.m., in room 2154, Rayburn House Office Building, Hon. Tom Davis (chairman of the committee) presiding.

Present: Representatives Tom Davis, Platts, Miller, Marchant, Dent, Schmidt, Waxman, Cummings, and Van Hollen.

Staff present: David Marin, staff director; Steve Castor, counsel; Chas Phillips, policy counsel; Rob White, press secretary; Victoria Proctor, senior professional staff member; Teresa Austin, chief clerk; Sarah D'Orsie, deputy clerk; Phil Barnett, minority staff director/chief counsel; Michael McCarthy, minority counsel; Earley Green, minority chief clerk; and Jean Gosa, minority assistant clerk.

Chairman TOM DAVIS. The committee will come to order. Good morning. Welcome. A quorum being present, the committee will come to order. I would like to welcome everybody to today's hearing on information sharing and the situational awareness during the management of an emergency. The purpose of this hearing is to reignite public discussion and debate on barriers to information sharing among agencies and highlight practices and procedures that could be effective in encouraging and enhancing information sharing among diverse entities.

The Government needs to be able to identify threats of all types and meet or defeat them. Our success depends on collecting, analyzing, and appropriately sharing information found in data bases, transactions, and other sources. Both the 9/11 Commission report and the Select Katrina Committee report made it clear there is a lack of effective information sharing and analysis among the relevant public and private sector entities.

We are still an analog Government in a digital age. We are woefully incapable of storing, moving, and accessing information, especially in times of crisis. Many of the problems in these times can be categorized as "information gaps"—or at least problems with information-related implications, or failures to act decisively because information was sketchy at best.

Unfortunately, no Government does these things well, especially big governments. The Federal Government is the largest purchaser of information technology in the world, by far, and one would think that we could share information by now.

The 9/11 Commission found "the most important failure was one of imagination." Katrina was primarily a failure of initiative. But there is, of course, a nexus between the two. Both imagination and initiative—in other words, leadership—require good information. And a coordinated process for sharing it. And a willingness to use information—however imperfect or incomplete—to fuel action.

With Katrina, the reasons reliable information did not reach more people more quickly were many, for example: the lack of communication and situational awareness paralyzed command and control; DHS and the States had difficulty coordinating with each other, which slowed the response; DOD lacked an information sharing protocol that would have enhanced joint situational awareness and communication between all military components.

Information sharing and situational awareness will always be predicated to an effective disaster response. With approximately 60 days remaining before the start of hurricane season on June 1st, this hearing will examine how the lessons learned regarding information sharing in the context of law enforcement, counterterrorism, and defense can be applied to disaster response.

Information sharing is the backbone of successful emergency preparation and response efforts. Historically, however, the Federal Government has been so compartmentalized, information sharing has been a pipe dream. The Federal Government is faced with the difficult task of transforming from a "need-to-know" information sharing environment to a "need-to-share." In addition, the bureaucratic stovepipe arrangement in Federal agencies restricts the Government's flexibility to analyze information quickly, assess the need for services, and respond effectively in emergency situations.

Governmentwide information policy authority rests with the White House, in the Office of Management and Budget. I think the White House, through OMB, has a critical role in establishing and implementing policies and procedures for Federal information sharing. Whether we are discussing disaster management, counterterrorism, or law enforcement, overarching guidance and oversight to help Federal agencies establish a structure for partnering with one another and local and State organizations.

Given the lessons learned from Katrina, emergency managers and officials are obligated to the American people to produce a more nimble, effective, and robust response to predictable natural disasters. How can we avoid the inadequate information sharing and murky situational awareness that characterized the Government response to Katrina? Are impediments to more effective information sharing primarily technological, structural, cultural, or bureaucratic in nature?

The committee's hearing will include a review of the issues raised by the Select Committee Report. This hearing is not intended to review the facts surrounding Hurricane Katrina, but will use the disaster to highlight instances where collaboration and information sharing among agencies is lacking. In addition, the committee will explore the barriers to effective information sharing, learn what entities—including State, local, defense, intelligence, homeland security, and industry—are particularly adept at information sharing, and examine the models, policies, and methods which have proven successful. Finally, the committee is interested in learning about whether there is a need for additional legislation, guidance, procedures, or resources to facilitate the information sharing priorities outlined by the witnesses.

The committee views this hearing as a new beginning on the road to improving information sharing among Government agencies and between the public and private sectors. To this end, private sector stakeholders and other key agency personnel, including representatives from the Department of Homeland Security and the Office of the Director of National Intelligence, will be asked to testify at future hearings.

[The prepared statement of Chairman Tom Davis follows:]

Oversight Hearing

"The Need to Know: Information Sharing Lessons for Disaster Response."

Thursday, March 30, 2006
10:00 a.m.
Room 2154 Rayburn House Office Building

Opening Statement

Good morning and welcome. A quorum being present, the Committee on Government Reform will come to order. I would like to welcome everyone to today's hearing on information sharing and situational awareness during the management of an emergency. The purpose of this hearing is to reignite public discussion and debate on barriers to information sharing among agencies and highlight practices and procedures that could be effective in encouraging and enhancing information sharing among diverse entities.

The government needs to be able to identify threats of all types and meet or defeat them. Our success depends on collecting, analyzing, and appropriately sharing information found in databases, transactions, and other sources. Both the 9-11 Commission report and the Select Katrina Committee report made it clear there is a lack of effective information sharing and analysis among the relevant public and private sector parties.

We are still an analog government in a digital age. We are woefully incapable of storing, moving, and accessing information – especially in times of crisis. Many of the problems in these times can be categorized as "information gaps" – or at least problems with information-related implications, or failures to act decisively because information was sketchy at best.

Unfortunately, no government does these things well, especially big governments. The federal government is the largest purchaser of information technology in the world, by far, and one would think we could share information by now.

The 9/11 Commission found "the most important failure was one of imagination." Katrina was primarily a failure of initiative. But there is, of course, a nexus between the two. Both imagination and initiative – in other words, *leadership* – require good information. And a coordinated process for sharing it. And a willingness to use information – however imperfect or incomplete – to fuel action.

With Katrina, the reasons reliable information did not reach more people more quickly were many. For example: 1) the lack of communication and situational awareness paralyzed command and control; 2) DHS and the states had difficulty coordinating with each other, which slowed the response, and 3) DoD lacked an information sharing protocol that would have enhanced joint situational awareness and communication between all military components.

Information sharing and situational awareness will always be predicates to an effective disaster response. With approximately 60 days remaining before the start of hurricane season on June 1, this hearing will examine how the lessons learned regarding information sharing in the context of law enforcement, counter-terrorism, and defense can be applied to disaster response.

Information sharing is the backbone of successful emergency preparation and response efforts. Historically, however, the federal government has been so compartmentalized, information sharing has been a pipe dream. The federal government is faced with the difficult task of transforming from a "need-to-know" information sharing environment to a "need-to-share." In addition, the bureaucratic stovepipe arrangement in federal agencies restricts the government's flexibility to analyze information quickly, assess the need for services, and respond effectively in emergency situations.

Government-wide information policy authority rests with the White House, in the Office of Management and Budget. I think the White House, through OMB, has a critical role in establishing and implementing policies and procedures for Federal information sharing. Whether we are discussing disaster management, counter-terrorism, or law enforcement, overarching guidance and oversight to help federal agencies establish a structure for partnering with one another and local and state organizations.

Given the lessons learned from Katrina, emergency managers and officials are obligated to the American people to produce a more nimble, effective, and robust response to predictable natural disasters. How can we avoid the inadequate information sharing and murky situational awareness that characterized the governmental response to Katrina? Are impediments to more effective information sharing primarily technological, or structural, cultural, and bureaucratic in nature?

The Committee's hearing will include a review of the issues raised by the Select Committee Report. This hearing is not intended to review the facts surrounding Hurricane Katrina, but will use the disaster to highlight instances where collaboration and information sharing among agencies is lacking. In addition, the Committee will explore the barriers to effective information sharing, learn what entities—including state, local, defense, intelligence, homeland security, and industry--are particularly adept at information sharing; and examine the models, polices, and methods which have proven successful. Finally, the Committee is interested in learning about whether there is a need for additional legislation, guidance, procedures, or resources to facilitate the information sharing priorities outlined by the witnesses.

The Committee views this hearing as a new beginning on the road to improving information sharing among government agencies and between the public and private sectors. To this end, private sector stakeholders, and other key agency personnel, including representatives from the Department of Homeland Security and the Office of the Director of National Intelligence, will be asked to testify at future hearings.

Chairman TOM DAVIS. I would now recognize the distinguished ranking member, Mr. Waxman, for his opening statement.

Mr. WAXMAN. Thank you, Mr. Chairman, for holding this hearing to examine issues raised by the failed response to Hurricane Katrina. The report of the Select Committee on Hurricane Katrina identified widespread and serious problems with our Nation's disaster preparedness and response. The Government Reform Committee must take the next steps in finding solutions to these problems so that the Government can better help our citizens through the next disaster.

This hearing on how to improve information sharing during a disaster is a good first step for our committee to take. I hope we can continue to work together on oversight of the Department of Homeland Security and other Federal agencies to make sure that better communications procedures and technology are put into place.

Right now, across the river in Alexandria, admitted al Qaeda member Zacarias Moussaoui is on trial, facing the death penalty for his role in the September 11th attacks. As we all now know, Mr. Moussaoui was in custody weeks before September 11th. His attendance at flight school raised alarms among some experienced law enforcement and intelligence professionals about a possible hijacking plot. But as the 9/11 Commission documented, the Government never pulled together the various threads of information that could have detected the September 11th plot. Better information sharing was one of the key recommendations that the 9/11 Commission made.

Hurricane Katrina showed us that serious flaws remain in the Government's crisis prevention and response communications capabilities.

The Katrina investigation revealed that President Bush, Homeland Security Secretary Chertoff, and other top officials were unaware of the magnitude of the disaster facing New Orleans until Tuesday, August 30th, a day after the levees broke. They were unaware of this even though the first reports of levee breaches came as early as 8 a.m. on Monday, and the levee breaches were confirmed by late afternoon that day.

In fact, as late as 2 weeks after landfall, President Bush continued to insist that the levees had not breached until Tuesday and that there was a sense of relaxation at the White House on Monday night and Tuesday morning because he and other top officials believed that New Orleans had "dodged a bullet."

This was an inexcusable failure of the most senior officials in our Government to comprehend and act on urgent warnings and vital information.

The second problem causing a lack of information was technological. Katrina was such a powerful storm that it knocked out phone lines and radio towers throughout a three-State region, leaving local officials unable to communicate their needs to State and Federal officials who had the resources to help. Some of this was unavoidable. Any large enough disaster is bound to damage or destroy telecommunications infrastructure. But there are options, like a satellite phone, that could provide redundancy and allow communications when the regular system is down. Yet these were not in place.

I understand that we invited officials from the Department of Homeland Security to testify today, but they declined the invitation. DHS clearly has a primary responsibility for information sharing during disasters, and I hope that we will have another hearing where we can hear from representatives of the Department of Homeland Security.

I want to give my thanks to all the witnesses who did appear today before us, and I am looking forward to their testimony.

Thank you.

[The prepared statement of Hon. Henry A. Waxman follows:]

**Statement of Rep. Henry A. Waxman, Ranking Minority Member
Committee on Government Reform
Hearing on "The Need to Know: Information Sharing for Disaster
Response"**

March 30, 2006

Thank you, Mr. Chairman, for holding this hearing to examine issues raised by the failed response to Hurricane Katrina. The report of the Select Committee on Hurricane Katrina identified widespread and serious problems with our nation's disaster preparedness and response. The Government Reform Committee must take the next steps in finding solutions to these problems, so that the government can better help our citizens through the next disaster.

This hearing on how to improve information sharing during a disaster is a good first step for our Committee to take. I hope we can continue to work together on oversight of the Department of Homeland Security and other federal agencies to make sure that better communications procedures and technology are put into place.

Right now, across the river in Alexandria, admitted Al-Qaeda member Zacarias Moussaoui is on trial, facing the death penalty for his role in the 9/11 attacks. As we all now know, Mr. Moussaoui was in custody weeks before September 11. His attendance at flight school raised alarms among some experienced law enforcement and intelligence professionals about a possible hijacking plot. But as the 9/11 Commission documented, the government never pulled together the various threads of information that could have detected the 9/11 plot. Better information sharing was one of the key recommendations that the 9/11 Commission made.

Hurricane Katrina showed us that serious flaws remain in the government's crisis prevention and response communications capabilities.

The Katrina investigation revealed that President Bush, Homeland Security Secretary Chertoff, and other top officials were unaware of the magnitude of the disaster facing New Orleans until Tuesday, August 30 – a day after the levees broke. They were unaware of this even though the first reports of levee breaches came as early as 8:00 a.m. on Monday, and the levee breaches were confirmed by late afternoon that day.

In fact, as late as two weeks after landfall, President Bush continued to insist that the levees had not breached until Tuesday and that there was a sense of relaxation at the White House on Monday night and Tuesday morning because he and other top officials believed that New Orleans had "dodged a bullet."

This was an inexcusable failure of the most senior officials in our government to comprehend and act on urgent warnings and vital information.

The second problem causing a lack of information was technological. Katrina was such a powerful storm that it knocked out phone lines and radio towers throughout a three-state region, leaving local officials unable to communicate their needs to state and federal officials who had the resources to help. Some of this was unavoidable – any large enough disaster is bound to damage or destroy telecommunications infrastructure. But there are options, like satellite phones, that could provide redundancy and allow communications when the regular system is down. Yet these were not in place.

I understand that we invited officials from the Department of Homeland Security to testify today, but they declined the invitation. DHS clearly has a primary responsibility for information sharing during disasters, and I hope that we will have another hearing where we can hear from representatives from DHS.

Thank you to all the witnesses for your appearance before us today.

Chairman TOM DAVIS. Thank you.
Any other Members wish to make statements?
[No response.]
Chairman TOM DAVIS. Members will have 7 days to submit opening statements for the record.
[The prepared statement of Hon. Elijah E. Cummings follows:]

Opening Statement of

Representative Elijah E. Cummings, D-Maryland

Hearing Entitled: "The Need to Know: Information Sharing Lessons for Disaster Response."

Committee on Government Reform
U.S. House of Representatives
109th Congress

March 29, 2006

Mr. Chairman,

Thank you for calling this critically important hearing to assess obstacles to our nation's ability to share information in a disaster situation.

Today's hearing reflects an understanding that information sharing is central to upholding government's most basic of responsibilities in order to respond rapidly and competently to its citizens' needs for food, water, shelter, and safety in times of national peril. Tragically, the Hurricane Katrina fiasco is a clear example of what happens when all levels of government do too little -- to prepare, too little to share information, too little to combat the weight of indifference -- too well.

Make no mistake, the failures of government before, during, and after Hurricane Katrina left many Gulf Coast residents abandoned for four days without food or water, contributed to entire communities crumbling into a state of chaos, and played a major part in the needless deaths and suffering of countless Americans.

Hurricane Katrina also demonstrated in no uncertain terms that we have much work yet undone to strengthen our national preparedness to respond to threats of human design and acts of nature. In light of Katrina, it is vitally important that we not respond to natural disasters in a spirit of powerlessness. While we do not have control over nature, we do control the policy choices that determine our capacity to lessen the impact of nature's mighty blows.

One of those policy choices is the quality of the system we have in place to remove any bureaucratic obstacles that impede the flow of information. I was remarkably troubled to learn that bureaucratic barriers prevented President Bush, Secretary Chertoff, and other senior officials from obtaining reports of levee failures and widespread flooding in New Orleans until a full day after they transpired. The following day, when an "all hands on deck" attitude should have characterized the federal government's understanding of the severity of the situation, President Bush said he felt as though New Orleans had "dodged a bullet" and Secretary Chertoff, likewise relieved, attended an avian flu briefing.

Clearly when our nation's senior officials fail to receive coordinated and timely information in a disaster situation, they are incapable of making well-informed life and death decisions.

We should also recognize that communication technology breakdowns and impediments helped to stifle efforts by local officials to convey to state and federal entities the necessity of immediate assistance.

These efforts would have greatly facilitated the delivery of aid after the most devastating natural disaster in our nation's history. Moreover, I am concerned by the fact that after the unnecessary deaths of first responders on 9/11 because of interoperability problems, local responders in the midst of Katrina were still unable to communicate with one another. So too, am I deeply concerned that first responders were relegated to literally employing runners to run from location to location delivering messages.

In the modern age where information has the potential to travel around the world instantaneously, and where interoperability solutions exist, it is a scandal that more substantive progress on this issue has not been made over four years after 9/11. All levels of government must find the will and the resources to do whatever is necessary to address our nation's interoperability problem.

Mr. Chairman, we must do more than simply recognize the myriad of information sharing failures associated with Katrina, we must be solution driven and address them. This hearing is a step forward in that respect and a step forward in meeting the American people's expectations that their government will get it "right" when confronted with another national catastrophe.

I yield back the balance of my time and look forward to the testimony of today's witnesses.

Chairman TOM DAVIS. We will now recognize our first panel: Mr. Peter Verga, the Principal Deputy Assistant Secretary of Defense for Homeland Defense, U.S. Department of Defense; Dr. Linton Wells, the Principal Deputy Assistant Secretary of Defense, Networks and Information Integration, U.S. Department of Defense; and Mr. Vance Hitch, the CIO of the Department of Justice.

It is our policy that we swear you in before your testimony, so if you would just rise and raise your right hands.

[Witnesses sworn.]

Chairman TOM DAVIS. Thank you very much.

Mr. Verga, Dr. Wells, who wants to go first? OK. Dr. Wells, we will start with you and then go to Mr. Verga and then, Mr. Hitch, you will be cleanup. Thank you very much.

STATEMENTS OF LINTON WELLS II, PRINCIPAL DEPUTY ASSISTANT SECRETARY, NETWORKS AND INFORMATION INTEGRATION, U.S. DEPARTMENT OF DEFENSE; PETER F. VERGA, PRINCIPAL DEPUTY ASSISTANT SECRETARY FOR HOMELAND DEFENSE, U.S. DEPARTMENT OF DEFENSE; AND VANCE HITCH, CHIEF INFORMATION OFFICER, U.S. DEPARTMENT OF JUSTICE

STATEMENT OF LINTON WELLS

Dr. WELLS. Thank you, Chairman Davis, Ranking Member Waxman, and distinguished members of the committee, for inviting me here today to discuss this important topic. I would like to introduce Ms. Deb Filippi, the DOD Chief Information Officer's Information Sharing Executive. She is charged with strengthening our information sharing.

While the Department of Defense Chief Information Officer is responsible for information sharing within DOD and with our partners, since the specific focus of this hearing is on following up on the report on Hurricane Katrina, I would like to pass the microphone to Mr. Verga. I would like, however, to note that everything that we have learned about information sharing from humanitarian assistance in tsunami and Katrina, to stabilization and reconstruction operations in Afghanistan and Iraq teaches us that successful information sharing and collaboration is much more than just technology. It involves policies and procedures, social networks, organizational training, and as the chairman has noted, leadership. All of these must be co-evolved with the capabilities in order to achieve successful outcomes.

I have submitted written testimony. I would like it entered for the record. I look forward to working with the Congress and industry on this important topic. I am ready to answer your questions.

[The prepared statement of Dr. Wells follows:]

STATEMENT BY

DR. LINTON WELLS II
PRINCIPAL DEPUTY ASSISTANT SECRETARY OF DEFENSE
(NETWORKS AND INFORMATION INTEGRATION)

BEFORE THE

HOUSE COMMITTEE ON GOVERNMENT REFORM

ON

THE NEED TO KNOW: INFORMATION SHARING LESSONS FOR

DISASTER RESPONSE

MARCH 30, 2006

**NOT FOR PUBLICATION
UNTIL RELEASED BY THE
HOUSE COMMITTEE ON GOVERNMENT REFORM**

INTRODUCTION

Chairman Davis, Ranking Member Waxman, distinguished members of the Committee, thank you for inviting me here today to discuss the subject of information sharing lessons for disaster response. As the Principal Deputy Assistant Secretary of Defense for Networks and Information Integration (NII), I am representing Mr Grimes as the ASD(NII). This office recently designated a senior executive to serve as the Department's Information Sharing Executive.

I want to set the larger context for how information sharing and situational awareness are important not just to the Defense Department but to many processes the government is involved in. Warfare in the 21st Century, the core business process of the Defense Department, must be net-centric, meaning so well connected that well-trained professionals can self-synchronize their behavior with many others across vast distances, with devastating effect. Victory is dependent on discovering the enemy, accessing data, making decisions, and executing operations more rapidly and effectively than your adversary. Let me begin by saying that the communications and command and control (C2) lessons we are learning from the Federal, state, local, and commercial responses to Hurricane Katrina appear consistent with the lessons DoD has learned in the conduct of Humanitarian Assistance and Disaster Relief missions across the globe. Moreover, these lessons appear consistent with those lessons learned during stabilization and reconstruction operations in Afghanistan and Iraq. All of these situations involve high-levels of complexity, large populations, and the destruction of basic information and communications infrastructure. There is also a commonality of purpose that must be organized, coordinated, deconflicted, and executed as efficiently and effectively as possible, using multiple sources of support – some of them totally unfamiliar with one another.

Communications – particularly wireless communications – are *the* critical enabler of all other functions in any disaster relief operation, along with the sensors to let you know what's happening and share the information and the ability to command and control those functions and information. These are all mission-critical functions. Hurricane Katrina was no exception. Without effective communications, every operation will suffer debilitating inefficiencies, some leading to ineffectiveness. My experience indicates that the first priority in both international and domestic situations is the establishment or restoration of wireless communications. Establishing or reestablishing communications has become a first-order requirement that must occur contemporaneously with rescue operations. Communication and information, when used appropriately, synergize the rescue response. It is imperative to take advantage of everyday technology to rapidly coordinate the rescue of our citizens across the entire spectrum of the crisis until its conclusion.

By now, the members of this Committee recognize that the Department of Defense and civilian responders from across the spectrum of Federal, state, and local authorities have matured into the post-September 11 world with different lexicons. The mission of fighting and winning this nation's wars is very different from responding to catastrophes spread across vast distances, regardless of their cause. Different lexicons are to be expected. America has a long tradition of carefully separating military and civilian functions, especially in our homeland. My experience, however, tells me that when Mr. Canterbury of the Fraternal Order of Police testified before the House Government Reform Subcommittee on Emergency Preparedness, Science and Technology last year, his reference to command and control is the same concept that General Pace, Chairman of the Joint Chiefs of Staff, refers to using the same words. The ability to lead a complex organized operation requires situational awareness and the ability to communicate with everyone participating in that operation. The planning process establishes the social networks and procedures that give people the agility to adapt and overcome the unanticipated.

CATEGORIZING CHANGE

From my experiences since September 11, I have come to use a three-part construct to describe the actions necessary to ensure operability in catastrophic events internationally and domestically. These categories include: 1) technical capacity development; 2) "social network" development through planning, interaction, and collaboration; and 3) doctrinal changes and training.

TECHNICAL CAPACITY DEVELOPMENT

During the past 10 years, the U.S. military has honed its C2 skills in multiple deployments involving a mixture of war-fighting, civil affairs, humanitarian assistance, disaster relief and stabilization and reconstruction operations. The 1990s saw such deployments in Haiti and the Balkans, and they have only accelerated since the 9-11 attacks, with deployments in Afghanistan and Iraq. More recently, U.S. forces have been instrumental in providing key elements of the initial humanitarian responses to global disasters, including the tsunami in Southeast Asia, the earthquake in Pakistan and Hurricane Katrina. All of these deployments have highlighted the increased need in the Department to communicate, collaborate, translate, and cooperate outside the closed networks required for military operations. Unlike the military, which always travels with its own power and infrastructure, civilian responders encountered command and control issues at the operational and tactical levels due to the devastation of the civilian-response infrastructure. Technology designed to operate without stable power sources in the austere environments of developing countries, is available today. Working with industry, these innovations can help to increase the survivability of tactical civil responder systems.

As stated earlier, when forces assigned to U.S. Northern Command and National Guard units deployed with military communications, they were once again ill-equipped to communicate with civilian responders struggling with a lack of communications infrastructure. Therefore, the Federal government must continue to expand its capability to rapidly deploy commercial-off-the-shelf networks making use of satellite links, wireless local area networks (LANs), laptop computers and "plug-and-play" equipment to bridge the gap created by a devastated civil infrastructure.

The lack of interoperability of first responders' communication equipment also hindered the effectiveness of operations. This problem won't be resolved by everyone buying the same product. It may be solved through collaborative efforts involving spectrum allocation and agreement both within industry and in the first responder community on common data standards. In the near term, we must continue to encourage the development and purchase of technology that bridge these disparate systems.

SOCIAL NETWORK DEVELOPMENT

Much of the work that needs to be done at the strategic level in the wake of what we have learned revolves around social networks rather than any lack of technology. Hurricane Katrina showed us that a key source of the problem stemmed from a lack of familiarity with each other's operating practices – what DoD calls tactics, techniques, and procedures. What was lacking was familiarity with the National Response Plan, a shared understanding of how NORTHCOM was to support that plan, and experience gained through exercises between U.S. military and Federal, state, and local responders. A nationally focused effort to generate a truly collaborative information environment is feasible through coordinating the resolution of legal, policy and technical issues across all agencies and all levels of government. Ideally, there would be full interoperability among systems for command and control, communications, computers, intelligence, surveillance, and reconnaissance (known together as "C4ISR"). In addition, there needs to be broader, more fully articulated planning for multiple kinds of disaster events, ranging from natural disasters such as Hurricane Katrina up through a nuclear strike. Command and control, which is a social process augmented by communications and information, must extend to all appropriate locations, from a local sheriff's car to the White House. Moreover, we must exercise and train in a common environment to be better prepared to respond to such crises in the future.

Multiple efforts have addressed, or are addressing, segments of the need for a national response capability. These include:

- National Security Telecommunications and Information Systems - Developing plans and programs, including the development of architectures, to ensure security on National Security Systems;

- Continuity Communications Enterprise Architecture – Architecture to enable the Federal Executive Branch to execute mission-essential functions under all circumstances;

- Intelligence Community Architecture – Architecture to enable the intelligence community to share information;

We must vigorously support collaborative planning and interoperability at all levels of government, ensuring that decision-makers have unencumbered access to the best available information and enabling interoperable command and control operations. The Federal government must have command and control capabilities, supporting facilities, and infrastructure to ensure uninterrupted connectivity and coordination in support of essential functions in accordance with constitutional authorities. Our goal should be to provide assured services across government by:

- Making information available on a network that is dependable and trusted,

- Providing the available and appropriate bandwidth, frequency and computing capabilities within the spectrum management process,

- Assuring appropriate and effective collaboration capabilities and other performance support tools,

- Supporting secure and assured information sharing, without disadvantaging the responder lacking a security clearance,

- Continuously refreshing the information content of a shared situational awareness capability,

- Promoting infrastructure transparency (to the user),

- Assuring independence of information and data for consumers and producers,

- Considering that all users of information are also suppliers (and therefore encouraging parties to contribute data rather than just downloading it),

- Supporting information transactions that are asynchronous in time and place,

- Supporting the disadvantaged user with intermittent access to limited data services, and

- Applying federal data tagging standards and information assurance policies.

I have learned a great deal about "social networks" in the international context in the past three years. It is critical to develop purposely professional and personal links among experts and practitioners from multiple fields and sectors in humanitarian relief, disaster relief, and stabilization and reconstruction operations. These ties, built up over time and through enormous effort, are absolutely vital to organizing an effective response when catastrophic disasters occur. Unless working arrangements to communicate and share information among all of these types of entities can be formulated, the success of any operation can be compromised, with results that can prolong or even exacerbate the effects of the disaster. Extensive planning and training is essential before the crisis.

DOCTRINAL CHANGES AND TRAINING

In the area of doctrinal change in the international context, DoD is embracing the concept of "integrated operations." This reflects a new battlespace management concept that will transform our military competencies from joint operations to operations that are fully integrated and coordinated with those of the military's partners in an operation. In the case of humanitarian assistance activities, these partners may include other U.S. agencies, allied militaries and governments, nongovernmental organizations, local populations, and private industry. And to maximize our effectiveness, DoD will integrate from planning to execution and then on to the transition to a restored local authority. Employing a coherent strategy that uses all instruments of the state in concert will ensure success in relief operations over the long term.

This doctrine also better prepares DoD to fulfill domestic response missions, bringing together civilian responders and military planners to synergize their efforts. Within the United States, DoD has conducted many scenario-driven exercises designed to prepare the military to support humanitarian assistance across a broad range of natural disasters – and also with regard to protecting potential terrorist target sites. Exercises and training opportunities between the U.S. military and civilian responders are critical to achieving this level of integration.

INFORMATION SHARING INITIATIVES

At this point, I would like to discuss some specific initiatives promoting the Information Sharing culture, a pilot effort with the Coast Guard, lessons learned with current techniques for regional information exchange, and a list of challenges we face as we prioritize and solve Inter-Agency infrastructure and information sharing efforts.

PROMOTING A CULTURE OF INFORMATION SHARING

The National Defense Strategy identifies "Conducting Network-Centric Operations" as one of the key operational capabilities required for defense transformation. The Department is transforming by building the foundation for net-centric operations in policy, program oversight, resource allocation, and cultural areas. Transforming to a net-centric force requires fundamental changes in all areas to provide the necessary speed, accuracy, and quality of decision-making critical to future success. Domestic response missions will benefit from this transformation and especially from two transformation initiatives promoting a culture of information sharing within DoD.

A cornerstone initiative is embodied in the DoD Net-Centric Data Strategy which outlines the vision for making data visible, available, and understandable--when needed and where needed--to accelerate decision cycles. The concepts in the Data Strategy are implemented through policies and actions carried out by Communities of Interest (COI) which are comprised of representatives from Combatant Commands, Services, Agencies and, where applicable, external partners. A COI is a community of people who are interested in the same subject and need to share information to resolve mission issues that affect their community.

The second initiative is achieving assured information sharing via transformation in Information Assurance (IA). As users, services, information, and networks are co-resident within a net-centric information sharing space, data encryption boundary protection requirements must now shift to each object, device, and user within DoD's Global Information Grid (GIG). Robust IA must be applied throughout the GIG to protect it against cyber warfare attacks. An assured enterprise requires enterprise level governance, systems engineering, policy, risk management, operational doctrine, and training properly integrated with technology. Achieving agreement on these aspects of an enterprise across the diverse set of DoD components, Services and Agencies, the Intelligence Community (IC), Department of Homeland Security (DHS), other Federal Agencies, state and local governments, tribes, non-government organizations, and our foreign partners is a major challenge. The IA component of the GIG Integrated Architecture, developed under the purview of the National Security Agency (NSA), provides an IA strategy for achieving the assured, integrated, and survivable information enterprise necessary to attain the strategic objectives of the DoD and IC. Additionally, DoD is focusing on some very specific information sharing initiatives.

INFORMATION SHARING ENVIRONMENT – EXECUTIVE ORDER 13388

Executive Order 13388, "Further Strengthening the Sharing of Terrorism Information to Protect Americans" makes clear the President's intent to ensure that the heads of all Federal departments and agencies who "possess or acquire terrorism information shall promptly give access to the terrorism information to the head of each other agency that has counterterrorism functions". Under the leadership of the Director of National Intelligence (DNI), DoD is fully engaged in supporting specific products such as establishing common standards for how information is acquired, accessed and used, addressing policy issues that inhibit information sharing and developing a common framework for information sharing between executive departments law enforcement agencies and state, local and tribal governments. An implementation plan is due in July of this year.

MARITIME DOMAIN AWARENESS (MDA) DATA SHARING PILOT

The Department of Defense (DoD), represented by US NORTHERN Command, in partnership with the Department of Homeland Security (DHS), represented by the United States Coast Guard, established a Maritime Domain Awareness Data Sharing Community of Interest (MDA DS COI) in February 2006. The MDA DS COI is piloting web-based data services to improve maritime situation awareness supporting Federal, State, Local, Tribal, Commercial, and international partners tracking vessels, cargo, and people of interest. This community-based pilot will develop a common vocabulary, and data services rendering maritime data visible, accessible, and understandable for authorized data users. The data users will post their data with appropriate context-related tags to improve the precision of subsequent data discovery and understanding. This effort is identifying and documenting a repeatable process to be applied to additional data sources in future pilot spirals for MDA and will be examined for applicability to other information sharing challenges. The MDA DS COI pilot goal is to demonstrate a methodology to increase maritime situational awareness and improve the security and defense of US borders and interests though the detection, tracking, interception, or interdiction of vessels, cargo, and people of interest within the maritime domain.

INFORMATION SHARING WITH OUR COALITION PARTNERS

In the execution of warfighting and stability operations, the ability to support combined operations is enhanced by sharing information among the participants. Currently, the primary network used by US Combatant Commanders to plan and execute operations is the Secret Internet Protocol Router Network or SIPRNET. Sharing information with coalition members in Afghanistan and Iraq requires three different networks. Information deemed releasable on the SIPRNET is downloaded onto separate media and transferred to be uploaded into the other networks.

The U.S. in collaboration with the United Kingdom, its key coalition partner, has established a senior body to bring greater information sharing capability to the warfighter quickly and be ready to fight together on the first day of the conflict. Mr. Grimes sits on this group, the U.S.-U.K. Interoperability Commission, along with Mr. Ken Krieg, the Under Secretary of Defense for Acquisition Technology and Logistics, and Lt Gen Fulton of the U.K. Our objective is to bring enhanced information sharing capability to the field today while ensuring a seamless transition to the future net-centric environment envisioned by the U.S. and U.K. The U.S. is supporting the U.K.'s upcoming

deployment into Afghanistan with real capability to share a common understanding of the

battle space. My team, working with the U.K., is implementing the capability to share

the Afghanistan common operational picture directly from the SIPRNET to the U.K.

Joint Operations Center system so it can provide the U.K. Commander with a robust

understanding of his operations area. This capability is expected to connect national

command and control systems directly, removing the burden of transferring information

from national systems to the appropriate network for sharing. These initial steps in a bi-

lateral environment are laying the groundwork for greater information sharing when

coalitions are involved. We are moving to provide greater combat effectiveness for U.S.

forces and the entire coalition.

INTER-AGENCY INFRASTRUCTURE WORKING GROUP

The DoD CIO office has established a working group with representatives from

many of the Federal Departments (DHS, State, DOJ) to identify and resolve issues that

inhibit the seamless exchange of relevant information. Some of the issues the working

group is addressing include: (1) governance (2) resourcing and prioritization of initiatives

(3) synchronizing of efforts (4) reconciling interdepartmental missions, strategies, and

objectives, and (5) trusted security accreditation.

In closing, allow me to reiterate that the road to effective information sharing and

situational awareness for the Department of Defense will be realized through our net

centric approach to data sharing and information assurance, our continued supremacy in state of the art communications equipment, our pursuit of the values that embrace a "sharing" culture and a commitment to changing our own doctrine and training structures.   This endeavor, however, is a journey...not a destination.

Thank you for the opportunity to address the Committee.

Chairman TOM DAVIS. Thank you very much.
Mr. Verga.

STATEMENT OF PETER F. VERGA

Mr. VERGA. Thank you, Mr. Chairman, and I appreciate the opportunity, along with the distinguished members of the committee, to come here to address today the Department of Defense information sharing lessons learned from disaster response.

Whether on the battlefield or in a disaster area, having the right information at the right time in order to take the right action can mean the difference between life or death, success or failure. DOD has a great deal of experience in the development and implementation of the essential policies, procedures, and technologies to enable effective information sharing and shared situational awareness.

That shared situational awareness—a common perception and understanding of the operational environment and its implications—is a core capability recognized in DOD's Strategy for Homeland Defense and Civil Support, which was published in June 2005.

The Quadrennial Defense Review, just recently completed, also recognizes the importance of shared situational awareness and calls for an information sharing strategy to guide operations with Federal, State, local, and coalition partners. The strategy for Homeland Defense and Civil Support supports this task and promotes the integration and sharing of applicable DOD capabilities, equipment, and technologies with Federal, State, local, and tribal authorities, and with the private sector.

While we are always striving to do better, DOD's approach to and capabilities for information sharing and shared situational awareness have proven effective over time. This performance is largely due to several organizational and cultural conditions within the Department.

First, DOD is a strategy-driven organization that plans for contingencies. Even as we marshal our currently available capabilities and resources to address a current situation, we are constantly planning and preparing for a full range of future contingencies.

As part of this planning culture, DOD expects and plans for complexity. We plan, for example, to deploy to and operate in regions where the supporting infrastructure, like roads, bridges, or communications, does not exist or has either been destroyed or seriously damaged.

Second, DOD has a highly disciplined yet flexible, multi-year focused budget and resourcing process that develops the capabilities necessary to deal with current and future contingencies.

And, third, as a military organization, DOD exercises unity of command over Federal military forces, DOD civilian personnel, and contractors at the strategic, operational, and tactical command echelons. This unity of command ensures both a unity of effort and an economy of force, that is, the right capabilities and forces in the right numbers.

Within the Department, DOD's command and control structure facilitates effective information flow between command echelons, whether the contingency is at home or abroad. When at home, a joint task force is established to command and control the Federal military forces, guided by the Commander of U.S. Northern Command in the joint operations area of a disaster. The NORTHCOM Commander in turn is responsible for ensuring that the joint task force receives the information it needs and provides information reported by the joint task force to the chairman of the Joint Chiefs of Staff and the Secretary of Defense.

Outside of DOD, several venues exist for information sharing between civilian and military and Federal, State, tribal, private sector, and nongovernmental organizations.

First, at the Federal headquarters level, incident information sharing, operational planning, and deployment of Federal resources are monitored by the Homeland Security Operations Center of the Department of Homeland Security, where DOD maintains a 24-hour-a-day/7-day-a-week presence. The HSOC, as it is known, facilitates interagency information sharing activities to enable the assessment, prevention, or resolution of a potential incident.

Second, strategic-level interagency incident management is facilitated by the Interagency Incident Management Group, which also serves as an advisory body to the Secretary of Homeland Security. When activated, the Department of Defense provides a senior-level representative to that IIMG.

Third, closer to the area of an incident, a Joint Field Office is established to provide a focal point for incident oversight and coordination of response and recovery actions. When established, the Department of Defense posts liaisons within the Joint Field Office known as Defense Coordinating Officers.

And, fourth, States usually maintain an Emergency Operations Center at which operational information sharing and resource coordination and support of on-scene efforts during a domestic incident activities normally take place during an incident and, when required, the Department will also deploy those Defense Coordinating Officers there.

Additionally, every combatant commander operates a Joint Interagency Coordination Group, which is a multi-functional, advisory element that represents the Federal civilian departments and agencies and facilitates information sharing. It provides regular, timely, and collaborative day-to-day working relationships between civilian and military operational planners.

Mr. Chairman, again, thank you very much for the opportunity to appear before you today. Thank you very much for the resources provided by the Congress and the American people to enable the Department of Defense to organize, train, and equip to meet the full range of DOD's missions, and I look forward to any questions that you may have.

[The prepared statement of Mr. Verga follows:]

Statement of

Peter F. Verga

Principal Deputy Assistant Secretary of Defense for Homeland Defense

Before the 109th Congress

Committee on Government Reform

United States House of Representatives

March 30, 2006

Introduction

Chairman Davis, Ranking Member Waxman, distinguished members of the Committee: thank you for the opportunity to address you today on the Department of Defense's (DoD's) information-sharing lessons learned for disaster response.

Whether on the battlefield or in a disaster area, having the right information at the right time to take the right action can mean the difference between life or death, success or failure. DoD has a great deal of experience in the development and implementation of the essential policies, procedures, and technologies to enable effective information-sharing and shared situational awareness among command echelons, the Military Departments, coalition partners, and non-governmental organizations (NGOs).

At home, the National Response Plan (NRP), which was published by the Department of Homeland Security in December 2004, recognizes the importance of information to mission success and states: "Implementation of the NRP and its supporting protocols will require extensive cooperation, collaboration, and information-sharing across jurisdictions, as well as between the government and the private sector at all levels." Providing mechanisms for vertical and horizontal coordination, communications, and information-sharing in response to threats or incidents is a key concept of the NRP. The mechanisms provided in the NRP are intended to facilitate coordination among State, local, and tribal entities, and the Federal Government, as well as between the public and private sectors.

Shared Situational Awareness

Shared situational awareness -- a common perception and understanding of the operational environment and its implications -- is a core capability indicated in

DoD's Strategy for Homeland Defense and Civil Support, which was published in June 2005.

Whether before, during, or after an incident, or at the level of senior decision-makers or the soldiers, sailors, airmen, or Marines actually carrying out decisions on-scene at an incident, shared situational awareness is a powerful enabler for mission success.

The Quadrennial Defense Review, which was published February 2006, tasked DoD to develop an information-sharing strategy to guide operations with Federal, State, local, and coalition partners. The Strategy for Homeland Defense and Civil Support supports this task and promotes the integration and sharing of applicable DoD capabilities, equipment, and technologies with Federal, State, local, and tribal authorities and the private sector.

Conditions for Success

While we are always striving to improve DoD's approach to and capabilities for information-sharing and shared situational awareness, our present capabilities have proven to be effective. This performance is largely due to several organizational and cultural conditions within DoD.

DoD is a strategy-driven organization that plans for contingencies. Even as we marshal our currently available capabilities and resources to address an emerging contingency, we are constantly planning and preparing for a full range of potential contingencies in the future.

DoD anticipates and plans for complexity. We plan, for example, to deploy to and operate in regions where the supporting infrastructure (e.g., communications, power, and roads) does not exist or has been destroyed or seriously damaged. We even plan for when things don't quite go as planned.

DoD has a highly disciplined yet flexible, multi-year focused budget and resourcing process that develops the capabilities necessary to deal effectively with current and future contingencies.

DoD plans campaigns, not just battles. Victory in war is often the accumulation of success over many battles, each planned to achieve specific objectives within the strategic context of an overall campaign, which is, in turn, driven by national objectives that direct and guide the exercise of the full spectrum of our nation's instruments of power.

DoD exercises unity of command over Federal military forces, DoD civilian personnel, and contractors at the strategic, operational, and tactical command echelons. This unity of command ensures a unity of effort and an economy of force (e.g., the right number of capabilities and forces) in all of DoD's actions and enables timely and effective decision-making at all command echelons. By design, unity of effort, not unity of command, is possible among Federal, State, and local, nongovernmental organizations (NGOs), and the private sector.

Incident Information-Sharing and Shared Situational Awareness

<u>Internal</u>

Within the Department, DoD's command and control structure, supported by a formalized reporting system and agile, resilient, and redundant communications, facilitates effective information flow between command echelons whether the contingency is at home or abroad. In the continental United States, a joint task force is established to command and control Federal military forces, guided by the Commander, U.S. Northern Command (USNORTHCOM). This joint task force is responsible for (a) ensuring that the tactical units under its command receive the information they need to fulfill their missions, and (b) ensuring that information, usually provided in the form of situation reports, is provided upwards to the Commander, USNORTHCOM. The Commander, USNORTHCOM, in turn, is responsible for (a) ensuring that the joint task force under its command receives the information it needs to fulfill its mission, and (b) providing information reported by the joint task force upwards to the Chairman of the Joint Chiefs of Staff and the Secretary of Defense.

<u>External</u>

Several entities exist for information-sharing among civilian and military and Federal, State, tribal, private-sector, and NGOs.

At the Federal headquarters level, incident information-sharing, operational planning, and deployment of Federal resources are coordinated by the Homeland Security Operations Center (HSOC). DoD maintains a 24 hours-a-day/7 days-a-week presence in the HSOC. At the national level, the HSOC facilitates interagency information-sharing activities to enable the assessment, prevention, or resolution of a potential incident. Federal, State, tribal, private-sector, and NGO emergency operations centers (EOCs) are either required or encouraged to report incident information to the HSOC. Federal departments and agencies are required to report information relating to actual or potential Incidents of National Significance to the HSOC. This information may include:

- Implementation of a Federal department or agency emergency response plan;

- Actions to prevent, respond to, or recover from an Incident of National Significance for which a Federal department or agency has responsibility under law or directive;

- Submission of requests for assistance to, or receipt of a request from another Federal department or agency in the context of an Incident of National Significance; and,

- Receipt of requests for assistance from State, local, or tribal governments; NGOs or the private sector in the context of an Incident of National Significance.

Strategic-level interagency incident management coordination and course-of-action development are facilitated by the Interagency Incident Management Group (IIMG), which also serves as an advisory body to the Secretary of Homeland Security. When activated, DoD provides senior-level representatives to the IIMG.

The Joint Field Office (JFO) is established in the area of the incident to provide a central point for Federal, State, local, and tribal executives with responsibility for incident oversight, direction, and/or assistance to coordinate protection, prevention, preparedness, response, and recovery actions effectively. Co-located within the JFO are all the entities essential to incident management, information-sharing, and the delivery of disaster assistance and other support. In the event that collocation is not practical, Federal agencies are connected virtually to the JFO and assign liaisons to the JFO to facilitate the coordination of Federal incident management and assistance efforts. When established, DoD posts liaisons within the JFO known as Defense Coordinating Officers (DCOs).

All states maintain an EOC in which operational information-sharing and resource coordination in support of on-scene efforts during domestic incident management activities normally take place. During an incident, DoD also posts DCOs within an affected State's EOC.

Additionally, each combatant commander, including U.S. Northern Command, operates a Joint Interagency Coordination Group (JIACG). The JIACG is a multi-functional, advisory element that represents the Federal civilian departments and agencies and facilitates information sharing across the interagency community. It provides an environmental for regular, timely, and collaborative day-to-day working relationships to develop between civilian and military operational planners.

Hurricane Katrina Lessons Learned

The Department of Defense response to Hurricane Katrina was the largest, fastest deployment of military forces for a civil support mission in our nation's history.

Hurricane Katrina made landfall along the Gulf Coast during the early morning hours of August 29th. Within five days of landfall, more than 34,000 military forces had been deployed into the affected area; that is more than five times the number of military personnel deployed within the same time frame in response to 1992's Hurricane Andrew.

By landfall plus seven days, more than 53,000 military personnel had been deployed in response to Katrina, three times the comparable response to Hurricane Andrew. And by September 10th, military forces reached their peak at nearly 72,000; 50,000 National Guardsmen and 22,000 active duty personnel, a total deployment for Hurricane Katrina, more than twice the size of the military response to Hurricane Andrew.

In scope and speed, no civil support mission in the history of the United States remotely approaches the DoD response to Hurricane Katrina.

DoD received 93 mission assignments from FEMA and approved all of them, on average, within 24 hours.

In addition to the 72,000 men and women in uniform, DoD coordinated the deployment of 293 medium- and heavy-lift helicopters, 68 airplanes, 23 U.S. Navy ships, 13 mortuary affairs teams, and two standing joint headquarters to support the Federal Emergency Management Agency's (FEMA's) planning efforts.

DoD military personnel evacuated more than 80,000 Gulf Coast residents and rescued another 15,000. Military forces provided significant medical assistance, including 10,000 medical evacuations by ground and air, the delivery of medical treatment to more than 5,000 sick and injured persons, as well as support for disease prevention and control.

DoD committed more than 2,000 health care professionals for civil support contingencies and approved six bases as FEMA staging areas. When violence erupted in New Orleans, Lieutenant General Blum, chief of the National Guard Bureau, coordinated, over a three-day period, the deployment of 4,200 National Guard military police and security personnel into New Orleans, dramatically increasing the security presence.

The President deployed 7,200 active duty military personnel for humanitarian relief; their presence, in combination with National Guard security forces, restored civil order in the city of New Orleans.

DoD delivered critical emergency supplies, more than 30 million meals, including 24.5 million meals-ready-to-eat and some 10,000 truckloads of ice and water.

Our performance, however, was not without defect. Hurricane Katrina and the subsequent sustained flooding of New Orleans exposed significant deficiencies in our national preparedness for catastrophic events and our Nation's capacity to respond to them. Emergency plans at all levels of government were put to an extreme test and came up short. As a result, President Bush, in his September 15, 2005, address to the Nation from Jackson Square in New Orleans, made it clear that the Federal government will make the necessary changes to be "better prepared for any challenge of nature, or act of evil men, that could threaten our people."

President Bush subsequently ordered a comprehensive review of the Federal response to Hurricane Katrina. This review resulted in the publication of "The Federal Response to Hurricane Katrina: Lessons Learned." Regarding DoD, the review states:

> The Federal response to Hurricane Katrina demonstrates that the Department of Defense (DoD) has the capability to play a critical role in the Nation's response to catastrophic events. During the Katrina response, DoD – both National Guard and active duty forces – demonstrated that along with the Coast Guard it was one of the only Federal departments that possessed real operational capabilities to translate Presidential decisions into prompt, effective action on the ground. In addition to possessing operational personnel in large numbers that have been trained and equipped for their missions, DoD brought robust communications infrastructure, logistics, and planning capabilities. Since DoD, first and foremost, has its critical overseas mission, the solution to improving the Federal response to future catastrophes cannot simply be "let the Department of Defense do it." Yet DoD capabilities must be better identified and integrated into the Nation's response plans.

These recommendations correlate well with DoD's own internal lessons learned effort. We have already begun to implement improvements with an urgent focus on the operational challenges associated with this year's upcoming hurricane season. For example, DoD is:

- Detailing a strategic planner to the DHS, and reviewing DoD personnel support to DHS in terms of both numbers and expertise to identify appropriate adjustments;

- Developing a framework to provide initial damage reconnaissance, including those capabilities provided by DoD organizations such as the National Geospatial Intelligence Agency (NGA) as part of a U.S. Strategic Command civil support plan;

- Participating in the interagency revision of the National Search and Rescue Plan, including disaster response operations and address air traffic control and coordination; and

- Pursuing better integration of Federal military forces and State National Guard forces, including greater shared visibility between the two on their respective deployments, maneuvers, and activities.

Conclusion

We recognize that the defense of the U.S. homeland – our people, property, and freedom – is our Department's most fundamental duty. Men and women in military uniform – Active Duty, Reserve, and National Guard – will continue to meet that obligation with passion, professionalism, and a resolute sense of purpose.

We also recognize that our Department has played and will continue to play an important supporting role in responding to natural or man-made disasters.

The ability of our military forces - Active Duty, Reserve, and National Guard – to respond quickly and effectively to an event on the scale of Hurricane

Katrina and to sustain simultaneously the ongoing War on Terrorism is a testament to their readiness, agility, and professionalism. It is also a reflection of the resources provided by Congress that enables them to organize, train, and equip to meet the full range of DoD's missions.

We have supported -- and expect to continue supporting -- national efforts to improve information-sharing and shared situational awareness, applying the experience, expertise, and technologies that we have developed over many years working abroad with Federal, coalition, and non-governmental partners.

Chairman TOM DAVIS. Thank you.

Mr. Hitch, thanks for being with us.

STATEMENT OF VANCE HITCH

Mr. HITCH. Good morning and thank you, Mr. Chairman and members of the committee, for the invitation to speak to you today. I am the Chief Information Officer of the Department of Justice, and next month will mark my 4-year anniversary with the Department. Today I will testify about our approach to information sharing.

The Department of Justice is committed to helping improve the ability of law enforcement and homeland security first responders to share national security information. This may include classified intelligence reports, criminal history records, or traffic stops. The key to all of this, though, is the data, helping over 180,000 law enforcement personnel follow standards so that they can safely and securely share photos, field reports, and evidence with a fellow officer.

First, I will focus on our umbrella program, the Law Enforcement Information Sharing Program. This program includes both internal DOJ sharing, such as between the Drug Enforcement Agency and ATF, and the Federal sharing with State and local law enforcement agencies and officers across the country.

The LEISP strategy is the result of a collaborative process including senior leadership from DOJ component agencies and representatives from across the national law enforcement community. LEISP is a program, not an information system. It addresses barriers to information sharing and creates a forum for collaboration on how existing and planned systems will be conducted and coordinated in a unified manner for information sharing purposes. LEISP delineates guiding principles, a policy framework, and functional requirements that are necessary to facilitate multi-jurisdictional law enforcement information sharing. LEISP establishes the Department's commitment to move from a culture of "need to know" toward a culture of "need to share" in which information is shared as a matter of standard operating procedure.

With our partners at DHS and the Department of Defense, we are making great strides in sharing fingerprints across boundaries. What we refer to as the Interoperability program is showing great returns as fingerprints captured in theater in Iraq are being sent to the FBI in West Virginia for comparison and coordination. DHS, under the US-VISIT program, has access to this data, and all three agencies are working on new standards to make this sharing even more timely and efficient.

As this committee is analyzing post-Katrina issues, I thought it was appropriate to mention two of the successes we had in the time immediately following the hurricane. As the Marshals Service moved prisoners from the New Orleans area, they faced the challenge of coordinating buses and new prison space. To complicate matters, the prisoners switched arm bands in hopes of confusing their guards. The Marshals used online photos and other descriptive data, such as scars, marks, and tattoos, from the joint automated booking system to ensure that valid identities were maintained. Another success story was the development and implementation of the National Sex Offender Public Registry through the support of the Bureau of Justice Assistance. This Web site was invaluable to law enforcement as it helped cities like Houston and Baton Rouge identify known offenders who had evacuated to their city. While this Web site was limited to one type of criminal, we see this as a model for other systems under development.

Now I would like to address a key question. What are some of the keys to success that we have found in planning and developing systems that share information within the law enforcement community?

The first is shared management. It is needed to create a federation of trust within the information sharing community. For example, the Attorney General's Global Information Sharing Initiative has brought together national leaders and law enforcement to help us develop our LEISP strategy and programs. Likewise, the Criminal Justice Information System Advisory Policy Board [APB], provides ongoing governance and working groups to help us as we build and operate information sharing systems, including criminal histories, incident reporting, uniform crime reporting, and fingerprints. Both the Global group and the CJIS APB are comprised of numerous State and local stakeholders.

The second key to success is the development of standards, which is an area where the Federal Government is expected to provide leadership. Two examples are the Global Justice XML Data Model and the National Information Exchange Model. Groups such as Global are important for setting, communicating, and maintaining national standards and a common vocabulary.

The widespread availability and use of Web services and commercial technologies will improve information standards over time. The Federal Government can help promulgate these standards through incentives such as grant programs and targeted technical assistance.

In response to the next disaster, data must be accessible from many places via many methods of telecommunications. Web-based systems, as opposed to those tied to a personal computer, allow an evacuated law enforcement officer, like the New Orleans P.D., to relocate to a city such as Irvine, TX, and still have access to their data. As long as the system has adequate back-up and recovery capabilities, many will be able to complete their work from alternate work locations. Katrina was a not-so-subtle reminder to Government personnel of the importance of continuity of operations and proper planning.

In closing, I want this committee to understand that the law enforcement information is being shared broadly at a local and regional level. The Department of Justice, in partnership with many Federal agencies, is attempting to make critical information exchanges more effective, more efficient, and more secure for our customers across the United States. We have many efforts underway that are validating our approach and pushing new concepts so that law enforcement personnel no longer need to think about sharing but, rather, it comes naturally and they share as a matter of practice.

Thank you for your time this morning, and I will be happy to answer any questions that you may have.
[The prepared statement of Mr. Hitch follows:]

Department of Justice

STATEMENT

OF

**VANCE E. HITCH
CHIEF INFORMATION OFFICER
UNITED STATES DEPARTMENT OF JUSTICE**

BEFORE THE

**COMMITTEE ON GOVERNMENT REFORM
UNITED STATES HOUSE OF REPRESENTATIVES**

CONCERNING

**"THE NEED TO KNOW: INFORMATION SHARING LESSONS
FOR DISASTER RESPONSE"**

PRESENTED ON

MARCH 30, 2006

Good afternoon and thank you, Mr. Chairman and Members of the Committee, for the invitation to speak to you today. I am the Chief Information Officer for the Department of Justice. I am proud to discuss the accomplishments the Department has made in the area of information sharing as I approach my four-year anniversary with the Department.

The Department is committed to helping improve the ability of law enforcement personnel and homeland security first responders to share national security information. This may include classified intelligence reports, criminal history records or traffic stops. The key to all of this is *data*, helping over 180,000 law enforcement personnel follow standards so that they can safely and securely share photos, field reports or evidence with a fellow officer. The Department has several ongoing programs that are designed to address particular aspects of the information sharing challenge. These include the National Data Exchange (N-DEx), Next Generation Identification (NGI) System and Regional Data Exchange (R-DEx). However today, I am focusing on our umbrella program, the Law Enforcement Information Sharing Program (LEISP). This program includes how Federal agencies share with each other, how internal Department components share with each other (*i.e.*, DEA to ATF), and most importantly, Federal sharing to local Law Enforcement (LE) agencies and officers.

The LEISP strategy is the result of a collaborative process involving senior leadership from the Department's component agencies and representatives from across the national law enforcement community. LEISP is a program, not an information "system." It addresses barriers to information sharing and creates a forum for collaboration on how existing and planned systems will be coordinated and unified for information sharing purposes. LEISP delineates guiding principles, a policy framework and functional requirements that are necessary to facilitate multi-jurisdictional law enforcement information sharing. LEISP establishes the Department's commitment to move from a culture of "need to know" toward a culture of "need to share" in which information is shared as a matter of standard operating procedure.

DOJ/DHS/DOD Inter-Operability

With our partners at DHS and the Department of Defense, we are also making great strides in sharing fingerprints across boundaries. What we refer to as the Inter-Operability program is showing great returns as fingerprints captured in theater in Iraq are being sent to the FBI in West Virginia for comparison and coordination. DHS (under the US VISIT program) has access to this data as well, and all three agencies are working on new standards to make this sharing even more efficient and timely.

Integrated Wireless Network (IWN)

I also want to mention the progress we have made on the Integrated Wireless Network (IWN) program. IWN is a partnership between the Department of Justice, the Department of Homeland Security and the Treasury Department to implement a consolidated nationwide communications system for federal law enforcement and homeland defense agents and officers. IWN will support approximately 80,000 Federal agents in all 50 States and the U.S. territories.

I testified to the House Committee on Commerce and Energy, Subcommittee on Telecommunications and the Internet in September of last year on the IWN program and we continue to move forward with the effort. We are nearly complete with the procurement and an award is expected soon.

<u>US Marshals Service and Sex Offender Website Success Stories</u>

As this Committee is analyzing post-Katrina issues, I thought it was appropriate to mention two successes we had in the time immediately following the hurricane. As the Marshals Service moved prisoners from the New Orleans area, they faced the challenge of coordinating buses and new prison space. To complicate matters, the prisoners switched arm bands in hopes of confusing their guards. The Marshals used online photos and other descriptive data (scars, marks and tattoos) to ensure that valid identities were maintained. Another success was the use of the National Sex Offender Public Registry (www.nsopr.gov). This website was invaluable to local law enforcement as it helped cities like Houston and Baton Rouge identify known offenders who had evacuated to their city. While this website was limited to one type of criminal, we see this as a model for some of our other systems under development.

<u>The Foundation is in Place</u>

Law enforcement agencies have been collecting and sharing information for decades. To support law enforcement needs, the Department and other law enforcement agencies have been providing actionable information that supports the mission and objectives of law enforcement agencies at all levels, by providing a variety of information sharing programs. The most far reaching effort is the Global Information Sharing Initiative (Global)[1]. Another example is the FBI's Criminal Justice Information System (CJIS), which provides law enforcement information relating to criminal histories, uniform crime reporting and fingerprint identification to meet the needs of

---

[1] Please see http://it.ojp.gov/topic.jsp?topic_id=8 for more information on Global.

Federal, State, local, and tribal law enforcement agencies. The State and local data providers and system users share responsibility for the operation and management of CJIS with the FBI through the CJIS Advisory Policy Board (CJIS APB). This shared management approach has provided the blueprint for the beginning of one of the most important prerequisites of successful information sharing: a federation of trust among all parties in the information sharing community.

Significant progress has been made to extend information sharing capabilities to a broader segment of the law enforcement community, and to begin connecting those disparate capabilities together. Far-reaching programs led by many agencies in the federal government have brought law enforcement partners at all levels together to address information sharing policy issues and develop standards such as the Global Justice XML (Extensible Markup Language) Data Model (GJXDM). These standards are developed in a public/private model with the help of non-profits and Federally Funded Research and Development Centers (FFRDCs). Groups such as Global are important for setting, communicating and maintaining national standards.

<u>Challenges</u>

The U.S. Constitution established a system of federalism, by which the responsibility for governing is divided between the national and the States' governments. However, it also means that law enforcement is organized into over 18,000 separate State, local, and tribal jurisdictions, with independent governance, information systems, and activities. The multiplicity of jurisdictions and their autonomous nature engender inconsistent policies, practices, and systems, and make coordination among agencies difficult. This also means that no one entity can mandate coordination across all agencies. We have found that in some instances like Southern California and the New York metropolitan area, the locals are ahead of the Federal agencies. However, they are still looking to the Federal government for standards, in order to expand the reach of their programs to neighboring jurisdictions. As we begin designing N-DEx, we also are looking to the private sector for tools and techniques for ingesting and indexing large data sets on an on-going basis. At the same time, we need to find ways to improve the quality of the data and protect the privacy rights of US citizens.

The Departments of Justice and Homeland Security have launched several specific IT projects to improve information sharing, including the effort to connect the Homeland Security Information Network (HSIN), Law Enforcement Online (LEO), Regional Information Sharing Systems Network (RISSNET), and the Criminal Intelligence Sharing Alliance Network (CISAnet). Finally, working with partners from other Federal, state, local and tribal agencies, as well as private industry, the Department has developed initial versions of information sharing technology standards for inter-system query and retrieval and for database inter-operability standards.

The Intelligence Reform and Terrorism Prevention Act of 2004 (IRTPA) and the President's Executive Order 13388 set out new requirements, mandates, and provisions for creation of an "Information Sharing Environment" (ISE), which will support the sharing of information among multiple communities, including law enforcement, intelligence, military, homeland security, diplomatic, State and local, and the private sector. I am personally involved with the ISE development on a weekly basis and many

Chairman TOM DAVIS. Well, thank you very much.

Let me start. This may not be a question you want to answer here. It is really to all of you. But information sharing, is this an issue that just cannot be overcome given agency structures and the congressional authorization and the appropriation process? I mean, we do things here to basically create stovepipes, too, just the way that we authorize, the way we appropriate. We have turf battles up here over the way committees operate. How would you suggest dealing with stovepipes, oversight, and funding? And how does that get into the mix of getting greater information sharing? Does anybody want to take a stab at that? Dr. Wells.

Dr. WELLS. I will start and I will pass to my colleagues. Since information sharing is a human activity, there are certainly going to be cultural and organizational biases that have to be addressed in the process of doing it. I would actually say that I think the cultural issues are probably significantly more important than the technical issues, given where we are today.

One of the things that the Department of Defense has done over the past several years is to do a series of demonstrations that we have called Strong Angel, and they have looked at not only the capabilities but all the sociological and, for us the military, doctrinal issues needed to overcome some of the information sharing.

One of the things in tsunami, for example, that we learned which applied to Katrina was we sent some people down there—a military doctor, a civilian doctor, and a retired Navy pilot—and what happened when they got to Jakarta is the two doctors were welcomed with abrazos by the nongovernmental organizations there because they had experience working together in Kosovo and Africa and places like this. The Navy pilot could go on board the carrier "Lincoln" and fit right into the aviation community.

What they found a few days later when they got together was that the military was prohibited by policy from sharing information outside the military boundaries unless asked. The nongovernmental organizations didn't know they had to ask and didn't know how to ask. Once those two groups got together, they were able to make enormous progress very quickly in sharing information. It was an issue of policy and procedures, not one of technology.

We applied some of this to Katrina, and there is an extensive exercise program that Northern Command is working on in preparation for the summer hurricane season to do this as well, to not only deal with the technologies but also bring together those groups of people that need to be able to cross these boundaries in communications. So I think that is at least as important a piece as any technological part.

Go ahead.

Mr. HITCH. Mr. Chairman, I have been in the Federal Government now for 4 years, and I would observe that the hardest things for me to accomplish had been to work across departments. So just the size of the organization is a barrier to communications. But I do think there are mechanisms in place that can make this more successful.

One that I would hold up as an example in the area of information sharing is the relatively recent identification within the DNI of the program manager's office, who is specifically chartered to come up with an information sharing environment, first to make sure that we are sharing terrorism information, but then more broadly to make sure that we're doing the things that we need to do to share information across Government departments. And this is something that I participate in on a weekly basis. We are having weekly cross-governmental meetings where we are actually on a very aggressive schedule to develop the concept of operations and the technology that is necessary to make sure that we are sharing information successfully.

The program that I mentioned that we have at Justice, the Law Enforcement Information Sharing Program, is something that I think I can bring to that group, because we have tried to do the same thing within our own community in law enforcement, and the DNI is actually trying to accomplish the same thing across Government.

I would say this is a good example because there does need to be a mechanism for bringing people together under some sort of—some appointed group who has a leadership authority, and that is the case of the program manager. So I think that is a good example.

In the case of emergency response and so forth, you know, the Department of Justice is not primarily a first responder organization, but in the Katrina situation, we did operate pretty effectively in our own community, within the law enforcement community. And I think that we share some traits with our DOD brethren in terms of, you know, having a command-and-control structure that is fairly regimented within the law enforcement community, and also the idea of we know that emergencies are going to happen, so we plan for them and we practice them.

So I think those are some things that I observed in my time in Government that has been successful and I think were the reasons for some of the success that we had in responding as we did in Katrina.

Chairman TOM DAVIS. Do you think the appropriations process plays a negative role in this, the way we appropriate up here?

Mr. HITCH. I think certainly it makes it more difficult. The appropriations process is a challenge for us as individuals trying to get support for the important programs that we are pursuing. But, once again, I am hoping—this is a little bit more hope than experience—that through the DNI, that will be a help in making sure that we get the support across appropriations groups and that somehow that will get the message across the line.

Personally, I have had reasonable success. In working with our appropriators, I think they understand the importance of the programs for information sharing and how important that is to us, not only for emergencies like Katrina but for, you know, responding to the counterterrorism challenge that we had after September 11th, and a lot of our programs are focused in that way. And so I think our appropriators have been reasonably responsible and responsive in helping me deal with those issues.

Chairman TOM DAVIS. OK. Thank you.

Do you have any comment on the appropriation process, either of you?

Mr. VERGA. I would only add one comment to what my colleagues have said, and that is that when we talk about information sharing as a problem to be overcome, it is good to keep in mind that it is one of those problems that does not have an end state that you can finally reach. There will always be more information to be shared than there are mechanisms for sharing it. And so I think the fact that we have made significant progress over the last 3 or 4 years I think shows us that progress can be made, but I don't know that we will ever reach an end state where people will be satisfied that all the information is being shared to the degree they would like to have it shared.

Chairman TOM DAVIS. Yes, Dr. Wells?

Dr. WELLS. Two things to go with it.

First of all, as we share information, there is such a thing as too much information, and one can—I have heard people complain now that so much information is being shared that they are drowning in data and that, if you will, the signal-to-noise ratio of valuable information to just useless makes it hard for them to find the nuggets. And so I think it is important to not only share, but share what is important for the problem at hand.

Chairman TOM DAVIS. OK. Thank you.

Mr. Cummings.

Mr. CUMMINGS. Thank you very much, Mr. Chairman. I want to thank you all for your testimony.

As I see it, we have two general categories of failure to communicate. You have the one where, I guess, agencies are not communicating properly. Then you have another one with regard—when you look at Katrina, with regard to communications equipment. And I got to tell you, when I read that back a few years ago, back when there was the Oklahoma bombing back in 1995, we had communication equipment problems. It is almost shocking to the conscience that we could come all the way up to 2005, in the greatest country, in the most powerful country, and one of the most technologically advanced countries in the world—in the world—and still have those kinds of problems.

It is interesting to note that when the folks from my State, Maryland, went down to the Gulf Coast, they discovered, Mr. Chairman, that they had better equipment and were better able to communicate than the FEMA folk, which was incredible to me. So that tells me that apparently the equipment is out there. The question is, you know, whether there are standards for communication equipment. In other words, I understand they were on different frequencies and all that kind of thing.

But I think that the thing that bothers me as I listened to all the testimony this morning is I wonder if we will be right back here 10 years from now, in other words, whether we will be saying the same things. Other problems will have occurred by then, and people will have died and people will have been in a position where, in a matter of less than, I guess, a 100-mile radius they cannot even communicate with each other.

So tell me, what are we doing with regard to equipment? What are we doing with regard to standards so that people can communicate? And keep in mind when you look at the data and you talk to the people in the Gulf Coast, you know what they said? They have said it over and over again. "We were not so much concerned about the fact that we had a disaster. We knew that those kind of things happen." They said that they felt abandoned as Americans, and part of that abandonment, I think, comes from the failure of us to be able to have simple communications, for me to be able to communicate across the street. And this is the United States.

I am just trying to figure out what are we doing about that. This is now our watch. Hopefully we have learned a lot from Katrina. I pray that we have. And if it is under our watch, what do we do from here? I mean, what are we able to say? What is on the drawing board? And what do you see correcting that communications problem?

Who is most appropriate to answer that? I guess you, Mr. Hitch?

Mr. HITCH. I don't know that I am the most appropriate, but I will take the first shot.

Mr. CUMMINGS. All right.

Mr. HITCH. I think, as I mentioned in my testimony, the standards issue is one that is being addressed and it is actually expected of the Federal Government. It is something that we are expected to do and we should be doing, and I think we are finally getting to the point where we are doing it. And I mentioned a couple examples in my testimony of some standards that have been kind of where the Department of Justice has taken the leadership role.

There is a program that I did not mention but that is in the written testimony called the IWN, Integrated Wireless Network. That is where law enforcement officers across the country still use radio communications. In the near term, in the not-too-distant future, it will be other forms of communication, but right now it is a lot of radios.

So we have a major program called IWN to set standards and to establish a nationwide network for law enforcement across the country. It is a cross-departmental effort between the Department of Justice in the lead, Department of Homeland Security, Department of Treasury, where between those three departments that is most of the law enforcement in the United States, to get them all on a common network with common equipment and common standards and all that kind of stuff. So that will help a lot.

So there are efforts, and on a local level, part of that program was something we called the 25 Cities Project, where we actually go in and kind of take each city and see what the problems are there and just try to help them solve them. In some cases, it was buying a piece of equipment. In some cases, it was providing training. So there are a lot of different things that cause communications problems that are not just all technical.

In terms of response to a disaster, in some of the examples that you alluded to where you could not communicate across the street and things like that, one of the things that I think are real lessons learned for CIOs like myself is the importance of back-up. Now, everybody for decades has known that you should have back-up for your information systems. But something like Katrina just brings that point home so clearly that the survivability of our systems are critical.

You know, as information systems have advanced and our workers have become more and more a part of their everyday life, we depend on them. So if they are without them for a period of time, they are at a loss. They can't do their job.

So taking those systems away and not having adequate back-up for those systems in time of emergency is just as bad as not giving them the system to begin with.

So the term "survivability" and how do we provide for that, and actually making sure that we are investing in the survivability of our systems is, I think, a real lesson learned.

In some cases where you—in Katrina, you were missing many layers of infrastructure and kinds of capabilities. You were missing the power. So if you were—electric power. So if your systems were all dependent on electric power and you had no back-up, battery back-up or anything else, you were out. In some cases, the back-up was gasoline-fired engines, and gasoline was not available either. So if your back-up depended on gasoline—first on electricity and then on gasoline, you were without. So it is really looking at what are the disasters that we are trying to address, what are the ones we have to plan for, and what kind of back-up is going to be needed in order to provide survivable systems under those circumstances. I think that is the biggest lesson from a technical standpoint.

The standards issues that you mentioned I think are real, and I think we are making a lot of progress in those areas.

Mr. CUMMINGS. Do we have any timetable for those standards? I mean particularly when you consider the fact that a lot of the same kinds of problems—if we had a terrorist attack, we would need the same kinds of communications systems or whatever. I mean, have you all set a timetable to try to have that done?

Mr. HITCH. Yes.

Mr. CUMMINGS. And as you were talking, I couldn't help but think about the fascination that my daughter, who is now a grown-up—I will never forget when she saw—you know, she said she couldn't believe that we were communicating, when she was a little girl, communicating on Earth to the Moon. To the Moon. She said, "Daddy, that's a joke." And then I think about how we are not even being able to communicate within a city, you know, it is just fascinating to me.

Mr. HITCH. Right. In the case that I gave you of the Integrated Wireless Network, we are embarking on a program that is going to take about, give or take a year or so, 5 years to get that rolled out across the country. And that will provide a long-term solution to the interoperability problem, but there are shorter-term solutions which we also have in the mix because we realize it is going to take a while to get it. There are technical solutions to solve the interoperability problem between different law enforcement organizations who happen to operate on a different standard. They are in existence today.

One of the things—again, back to Katrina, mobility is one of the things that is important. I mean, if all your infrastructure is out, being able to bring in something which is mobile on the back of a truck or something like that to power the equipment is something that I think really came home in that kind of an emergency situation.

But on the standards issue, as I said, I think standards by definition is a longer-term issue, longer-term solution to problems. It is the ultimate solution, but it is a longer-term solution. But there are shorter-term answers that we have to have in the mix also.

Chairman TOM DAVIS. Mrs. Schmidt.

Mrs. SCHMIDT. Thank you very much. I am not sure on the panel who should answer this, maybe all three of you. I think one of the biggest glitches with Hurricane Katrina was the inability for people all over the ground to communicate with each other, and I know that can be a local, a State, and a Federal issue.

From the Federal perspective, how can we coordinate communication so that the people on the ground know what is happening better. And I know that there are going to be some proposals later from people outside of Government talking about this very issue. What kind of sensitivity do we have from a governmental perspective of a security of information perspective? And how can we make the whole issue of communicating efficiently and effectively better, and better pretty soon? Because the next natural disaster or, God forbid, terrorist disaster could happen at a moment's notice.

Mr. VERGA. That, of course, is the nugget of what we are trying to do. One of the things that you have highlighted is that there is a fundamental difference between interoperability of communications, that is, existing communications being able to work together, and the operability of communications. What we discovered in Katrina was the issues were more on the basic operability of the communication side rather than the interoperability. There were interoperability issues, that is, system A and system B were not compatible, couldn't talk to each other. There are a lot of initiatives underway to fix that particular part of the problem.

But in a situation where you have the majority of the communications infrastructure, not just the public safety and security communications infrastructure, but the common infrastructure generally that is destroyed, the fundamental policy question is sort of what is the role of the Federal Government in this case in restoring those communications in a disaster area. Most of the communications are commercially owned, so how do you communicate with the American people? Hundreds of radio towers are down. Television stations are off the air. The normal means of communicating with the American people were not available.

So what, in fact, then is the role of the Federal Government in restoring that communications infrastructure in a disaster area?

Dr. WELLS. One of the initial proposals, for example, was that the Department of Defense or the Department of Homeland Security should stockpile radios that could be handed out in this type of emergency. Well, part of the problem is given the pace of technology today, if they have a warehouse full of radios that are degrading at the rate of Moore's Law, or whatever, it is not a very attractive way to do business.

There have now been a number of proposals to tap the genius of the private sector, especially for the nongovernmental, and so one example, for example, is use leased services that says I need to be able to have a certain amount of communications up and a certain amount of communications down at three spots anywhere in the United States within 12 hours. And, you know, we will keep you on retainer to be able to provide that capability, and maybe 10 spots in 72 hours.

So this type of approach gets the Government out of the business of warehousing equipment that could be obsolescent, allows for the continual upgrading of the capabilities, and involves the private sector more.

A related piece of this is that technology in this case is actually on our side because the Internet protocol, which is the basis of so much of our Internet communications, is now being able to be extended to mobile communications as well. And that then allows you to bridge lots of different incompatible systems, and I think that should be able to help.

If I may make one final point that the Congress could help with, the emergency responders, the keepers of critical infrastructure—power, water, telecommunications—are not now designated under the Stafford Act as emergency responders, and this got into problems in at least Wilma, I don't know about Katrina, but of people who wanted to go in and restore telecommunications, not being allowed through the security boundaries because they had no valid credential as an emergency responder. And so if there are ways to make adjustment to that, I think it could be a real term fix.

Mrs. SCHMIDT. Mr. Chairman, may I have a followup?

Chairman TOM DAVIS. Yes, go ahead.

Mrs. SCHMIDT. In followup to this, gleaning through future people that will be before us today, one of the things that came out of additional testimony is an apparent lack of leadership on the ground, who was really in charge. You talk about people that wanted to help and didn't have a clearance to help. Should we have a designated body at the Federal level that, when a disaster hits a community, whatever agency at the Federal level will be ultimately and automatically in charge so that you don't have the tension that may have been created on the ground between two competing agencies, maybe a State, maybe a local? And let me tell you where I am coming from. I know that in some cases, there are laws that are written in various States and in various communities that these local agencies have a certain jurisdiction. And it is not a turf battle of power. It is a turf battle of the way those local laws are written. And I don't think it is incumbent upon us to demand that those laws be rewritten, but I think it is incumbent upon Congress to figure out that in certain cases—a national emergency, a hurricane disaster at the level of Katrina—that somebody supersedes those locals on the ground so that we do not have this kind of confusion.

Having said that, how do you think that should be and who do you think should ultimately be the decisionmaker?

Mr. VERGA. You have addressed what is one of the fundamental challenges of federalism when you talk about how the Federal Government responds to any situation that is local in nature. The current national policy is, of course, that initial responsibility for responding to disasters of any type is within the local officials and then with the State officials. And that is embodied essentially in the Stafford Act as the legislation that talks to how we respond to disasters.

The legislation that established the Department of Homeland Security gave to that Department the responsibility for coordinating the national response to any type of emergency, natural disasters included. The principle that it operates under is one of unity of effort as opposed to unity of command, which is a term which is near and dear to the military. We always know who is in command of military forces. When you talk about organizing the efforts of everyone from a parish sheriff to the Federal Emergency Management Agency and the Department of Defense, that is a coordinating effort, not a command effort. The command, you know, on the Federal side comes together only at the President, and in the local side it depends on the State, how different States are organized—Commonwealths, States, those types of things.

My personal view is I am not sure there is, in fact, a legislative solution to that issue. The White House did an extensive study, as you are aware, which was recently published, on the lessons learned from Hurricane Katrina that talks to how we better organize the Federal effort to assist, but I don't think contemplates removing or superseding the authorities of State authorities beyond those provisions of the law which already exist. There are several provisions in the law that go back in history that allow, upon request of the State or, in the absence of a request, upon the determination of the President, that Federal authority needs to be asserted in a given jurisdiction, that can occur.

So I think the mechanisms are probably there. I think if there is something that Congress can do that can help, it's to assist in implementing those types of standards that make that process of getting the unity of effort to work better in terms of how moneys are appropriated, grants are given, those sorts of things.

Chairman TOM DAVIS. OK. Thank you very much. I have one other question.

Dr. Wells, in your written testimony, you discussed the need to establish social networks of Federal, State, and local partners as a critical component of successful response to catastrophic events. You state that one of the problems with the response to Hurricane Katrina was the lack of familiarity with each other's operating practices and experiences gained through exercises between the U.S. military and Federal, State, and local partners.

I think that is true. Although they had gone through the Hurricane Pam exercise at some time, you know, in an effort to try to get there, what efforts have you all taken to establish social networks? Can you describe briefly any exercises you have or plans you have with these partners?

Dr. WELLS. I mentioned earlier the Strong Angel series. There have been two of those that have expressly been looking at how, in the first case, military medicine reaches out to nongovernmental organizations in refugee situations; the second focused on Iraq and Afghanistan stabilization and reconstruction operations and sort of an Arab world type situation; a third this summer will focus—in August, will focus on an avian flu sort of situation, with more domestic, State and local responses.

Where this bore fruit was in tsunami, particularly, but also the group came together for Katrina, where we developed a virtual emergency operations center built around a commercial collaborative tool, and in there, there were over 600 people, and you could go in—who had sort of signed up. It was all voluntary. So you could say, "I need neurosurgeons who speak Bahasa Indonesia and also have had experience in southern Thailand," and find such people to go and work the problem. That group has sort of stayed virtually together and is available to be brought to bear on, you know, contingencies around the world, including domestic ones.

So it has been an ad hoc type of effort, but I think these types of—the only way you get the trust among these groups—I mentioned the case in Indonesia where the doctors could walk into the U.N. liaison center and be greeted because they were one of them. You cannot just say, "OK, you are in charge today and go bond with the people of New Orleans." If you have not built up those relations over time, it will be very hard.

So I think this is something we need to establish—to continue doing, and it will probably be regionally based. The people who would respond best in the Gulf Coast may be different than those who would go to San Diego in case of an earthquake. And so as we build this corps, we just need to understand the strengths and weaknesses and be able to mix and match on the fly, using information technology, to put together the best team for the situation required.

Chairman TOM DAVIS. Thank you very much.

I want to thank this panel. It has been very helpful for us. We appreciate the job that all of you are doing. The challenges remain ahead. So I will dismiss this panel and take about a 1-minute recess as we get our next panel.

Thank you very much.

[Recess.]

Chairman TOM DAVIS. We will recognize our second panel. We have John Brennan, president and CEO of the Analysis Corp. Thank you for being with us. Dr. Donald F. Kettl, the director of the Fels Institute of Government at the University of Pennsylvania. Dr. Brian Jackson, a physical scientist at the RAND Corp. And Lieutenant Steve Lambert, Virginia Fusion Center, Virginia State Police.

I want to thank all of you for being here. I am going to ask you to rise and raise your right hands.

[Witnesses sworn.]

Chairman TOM DAVIS. Thank you.

We will start, Mr. Brennan, with you. We may have a vote in about half an hour, and so I am going to try to get through. Once the bells ring for the vote, we will have about 10 minutes before I will have to go over to vote. But it will be our goal to try to finish up at that point and get you out of here. So if you can keep your testimony to 5 minutes, your total written statement is in the record, and my questions are based on having gone through that. Thank you very much. Mr. Brennan, you may start. And thanks again for being with us.

STATEMENTS OF JOHN BRENNAN, PRESIDENT AND CHIEF EXECUTIVE OFFICER, THE ANALYSIS CORP., McLEAN, VA; DONALD F. KETTL, DIRECTOR, FELS INSTITUTE OF GOVERNMENT, UNIVERSITY OF PENNSYLVANIA, PHILADELPHIA, PA; BRIAN A. JACKSON, PHYSICAL SCIENTIST, RAND CORP.; AND LIEUTENANT STEVE LAMBERT, VIRGINIA FUSION CENTER, VIRGINIA STATE POLICE

STATEMENT OF JOHN BRENNAN

Mr. BRENNAN. Good morning, Mr. Chairman. Thank you very much for the invitation to appear today. The views I offer today are my own, but they are informed by 25 years of experience as a CIA official as well as by my tenure as head of the Terrorist Threat Integration Center and of its successor organization, the National Counterterrorism Center.

The term "information sharing" has become one of the most frequently used phrases in Government since the devastating terrorist attacks that occurred on September 11, 2001. Members of Congress as well as senior officials in the executive branch have railed against the lack of sufficient sharing of critical information among Government agencies. The problem has been attributed, at various times, to institutional stovepipes, bureaucratic malaise, turf battles among agencies, excessive security requirements, mismanagement of IT resources and budgets, and a lack of strong and visionary leadership. I do not disagree that these factors have played a role in preventing the flow of relevant information in a timely fashion to departments, agencies, and individuals in need of such information.

But these factors have been allowed to flourish because of a much more fundamental systemic problem that afflicts our Government and our Nation in dealing with matters such as terrorism, hurricanes, a potential avian flu pandemic, or other challenges that may be on the horizon. The systemic problem is the absence of a coherent national framework that integrates and delineates roles and responsibilities on issues of major significance. Such a framework is the essential prerequisite to an effective information sharing regime that optimizes the formidable capabilities, knowledge, and expertise that are available in Federal, State, and local governments as well as in the private sector.

The purpose of sharing information is to ensure that individuals, departments, and organizations are able, in a timely fashion, to take some action or to perform some function for which they are responsible. Such actions and functions include warning and notification, protection and security, analysis and forecasting, rescue and recovery, policy decisionmaking, preparedness, and consequence management—just to name a few. The challenge for information providers, however, is that these diverse responsibilities are shared by many and are scattered across Federal, State, and local jurisdictions.

In the absence of an overarching framework, or "business architecture," that effectively integrates and articulates these responsibilities, the collectors, knowers, and stewards of relevant information are forced to make presumptive judgments about "who" needs access to "what." Similarly, the wanters of information are unsure to whom and to where they should look for information that addresses their needs. Confusion on both sides of the information divide has stymied the development of a symbiotic and synergistic relationship between information providers and users.

Unfortunately, it will take our Nation many years to adapt our outdated 20th century institutions, governance structures, and day-to-day business processes so that we may more effectively meet the challenges of the 21st century. In the meantime, and based on my experience setting up counter-terrorism organizations and information sharing practices across the Federal Government, I strongly recommend the establishment of a common information sharing and access environment that can be utilized by the providers and users of natural disaster information—whether they be Federal, State, or local officials, law enforcement agencies, the private sector, or U.S. persons seeking information so they can make appropriate decisions for themselves and for their families.

Specifically, I recommend the establishment of a Web-based portal on the Internet that would serve as a National Hurricane Information Center. Administered by the Federal Government, the portal would allow authorized information providers to post information and enable users to self-select information they need. Such a portal could serve as a one-stop shopping data mart containing virtually limitless archived and new information related to hurricanes, such as emergency contact information, weather reports, maps, first responder directories, hospital and health care providers, casualty and damage information, critical needs relief providers, security bulletins, shelter locations, and other relevant matters. Information could be organized and searched according to functional topics, geographic regions, or chronologically.

The portal could be constructed in a very flexible and versatile manner. In addition to providing general information to anyone who logs on as well as password-protected proprietary information accessible only to authorized users, the portal could serve as a communication mechanism among communities of interest, such as first responders. Unlike in the intelligence community, where complicated security requirements and multiple classified information networks inhibit the creation of a common information sharing environment, natural disaster information is not so encumbered. Thus, the ubiquity and robustness of the Internet makes it the ideal information sharing and information access platform for the Nation.

While the Federal Government would design and maintain the portal, there would need to be shared responsibility for posting, managing, and updating the content according to an agreed-upon business framework. The Federal Government also would have the responsibility for ensuring the portal's availability during emergencies and periods of peak activity and for the deployment of back-up systems when infrastructure is damaged. While this portal would not take the place of established information technology networks that serve as command-and-control mechanisms for individual departments and agencies, the portal would serve as a shared, collaborative information sharing and information access environment transcending individual entities.

Our Nation faces numerous challenges in the years ahead. In my view, confronting these challenges successfully hinges squarely on the Federal Government's ability to integrate capabilities and to leverage technology in an unprecedented manner within a national framework.

I look forward to taking your questions.

[The prepared statement of Mr. Brennan follows:]

"The Need to Know:
Information Sharing Lessons for Disaster Response"

****

House Committee on Government Reform Hearing
30 March 2006

Testimony by John O. Brennan

Good morning Mr. Chairman and distinguished members of the Committee. My name is John Brennan, and I am currently President and CEO of The Analysis Corporation of McLean, Virginia that provides analytic and technical support to the national counterterrorism community. It is a pleasure to appear before you today to address issues related to information sharing and situational awareness during the management of an emergency. The views I offer today are my own. They are informed by 25 years of experience as a CIA official as well as by my tenure as head of the Terrorist Threat Integration Center and of its successor organization the National Counterterrorism Center.

The term "information sharing" has become one of the most frequently used phrases in government since the devastating terrorist attacks that occurred on 11 September 2001. Members of Congress as well as senior officials in the Executive Branch have railed against the lack of sufficient sharing of critical information among government agencies. The problem has been attributed, at various time, to institutional stovepipes, bureaucratic malaise, turf battles among agencies, excessive security requirements, mismanagement of IT resources and budgets, and a lack of strong and visionary leadership. I do not disagree that these factors have played a role in preventing the flow of relevant information in a timely fashion to departments, agencies, and individuals in need of such information.

But these factors have been allowed to flourish because of a much more fundamental systemic problem that afflicts our government and our nation in dealing with matters such as terrorism, hurricanes, a potential Avian Flu Pandemic, or other challenges that may be on the horizon. The systemic problem is the absence of a coherent national framework that integrates and delineates roles and responsibilities on issues of major significance. Such a framework is the essential prerequisite to an effective information-sharing regimen that optimizes the formidable capabilities, knowledge, and expertise that are available in federal, state, and local governments as well as in the private sector.

Let me explain. The purpose of sharing information is to ensure that individuals, departments, and organizations are able, in a timely fashion, to take some action or to perform some function for which they are responsible. Such actions and functions include warning and notification; protection and security; analysis and forecasting; rescue and recovery; policy decision-making; preparedness; and consequence management; just to name a few. The challenge for information providers, however, is that these diverse responsibilities are shared by many and are scattered across federal, state, and local jurisdictions.

In the absence of an overarching framework, or "business architecture," that effectively integrates and articulates these responsibilities, the collectors, knowers, and stewards of relevant information are forced to make presumptive judgments about "who" needs access to "what." Similarly, the wanters of information are unsure to whom and to where they should look for information that addresses their needs. Confusion on both sides of the information divide has stymied the development of a symbiotic and synergistic relationship between information providers and users.

Unfortunately, it will take our nation many years to adapt outdated 20th Century institutions, governance structures, and day-to-day business processes so that we may more effectively meet the challenges of the 21st Century. In the meantime, and based on my experience setting up counterterrorism organizations and information-sharing practices across the Federal Government, I strongly recommend the establishment of a common information-sharing and access environment that can be utilized by the providers and users of natural disaster information—whether they be federal, state, or local officials; law enforcement agencies; the private sector; or U.S. persons seeking information so that they can make appropriate decisions for themselves and for their families.

Specifically, I recommend the establishment of web-based portal on the Internet that would serve as a "National Hurricane Information Center." Administered by the Federal Government, the portal would allow authorized information providers to post information and enable users to self-select information they need. Such a portal could serve as a one-stop shopping data mart containing virtually limitless archived and new information related to hurricanes, such as emergency contact information, weather reports, maps, first responder directories, hospital and health care providers, casualty and damage information, critical needs relief providers, security bulletins, shelter locations, and other relevant matters. Information could be organized and searched according to functional topics, geographic regions, or chronologically.

The portal could be constructed in a very flexible and versatile manner. In addition to providing general information to anyone who logs on as well as password-protected proprietary information accessible only to authorized users, the portal could serve as a communication mechanism among communities of interest, such as first responders. Unlike in the Intelligence Community, where complicated security requirements and multiple classified information networks inhibit the creation of a shared information-sharing environment, natural disaster information is not so encumbered. Thus, the ubiquity and robustness of the Internet makes it the ideal information-sharing and information-access platform for the nation.

While the Federal Government would design and maintain the portal, there would need to be shared responsibility for posting, managing, and updating the content according to an agreed-upon business framework. The Federal Government also would have the responsibility for ensuring the portal's availability during emergencies and periods of peak activity and for the deployment of back-up systems when infrastructure is damaged. While this portal would not take the place of established information technology networks that serve as command and control mechanisms for individual departments and agencies, the portal would serve as a shared, collaborative information-sharing and information-access environment transcending individual entities.

Our nation faces numerous challenges in the years ahead. In my view, confronting those challenges successfully hinges squarely on the Federal Government's ability to integrate capabilities and to leverage technology in an unprecedented manner within a national framework.

I look forward to responding to your questions.

Chairman TOM DAVIS. Thank you very much.
Dr. Kettl.

STATEMENT OF DONALD F. KETTL

Dr. KETTL. Mr. Chairman, thank you very much for the opportunity to appear before you this morning and to explore these issues about information sharing and operational awareness. The report by the Select Committee on Hurricane Katrina has already made an important national contribution to the question of how best to try to share information and to build a robust national system that could respond to the issues that we face.

The fundamental problem, however, is that we have too much thinking from the top down and not enough from the bottom up, and our principal goal needs, indeed must be, to create a system from the top down that works from the bottom up. That is the real driving meaning of what operational awareness means, to make sure that as we construct our systems, that it is real for the citizens who need help. And as the Select Committee identified, we have important issues about communication as well as command that we need to try to deal with.

The committee today has identified four basic questions that it wants to explore: culture, technology, structure, and bureaucracy. And as you sort through this, the thrust of both my testimony and of some of the other lessons that you have heard is how important the cultural piece is in establishing leadership and produce results.

The fundamental question here is what it is that we need to be focusing on. The focus so often on the cultural side is on a narrow stovepipe view of issues, but those issues and those structures never match the way the problems actually occur, whether on issues of terrorism or natural disasters. We need an all-hazard approach at the grass-roots level that will allow us to create a capacity for the Government to respond to the problems as they, in fact, arise.

The second thing is that we clearly have some technological issues that we need to face, in part making sure that we have communications systems that work in times of disaster and that connect with each other in times of disaster. I have talked with National Guard officials in Louisiana who have told me that one of the biggest problems that they had, even with people from the National Guard from around the country arriving to try to help, was that they arrived with radios that could not talk to each other, even within the National Guard. And those are issues that, Mr. Chairman, we fundamentally have to deal with.

We have some structural issues. If we had it do over again, we probably would not put FEMA inside the Department of Homeland Security, but we also know that continual disruption to FEMA's operations would only get in the way of getting the job done. The more fundamental issues are that we really cannot design any single structural solution that is guaranteed to solve whatever problem we face. The lesson of an all-hazards approach means that we must have a much more flexible and dynamic system that adapts our governmental operations and capacity to the problems that, in fact, we do confront.

One of the interesting things, in fact, is to look at FEMA's regional boundaries and compare that to the path of Hurricane Katrina, and Hurricane Katrina somehow miraculously found precisely the dividing line between the regions. There is no reason to think that if we were to redesign the regions that we would then not simply confront the same set of problems the next time we face an issue like this.

The last piece has to do with the bureaucracy, and it is clear that we have rules and procedures and other things that too often get in the way. What is also clear is that operational awareness teaches an important lesson, that if we focus on results to focus on outcomes, we can focus all those throughout the system on what it is that really matters most.

The good thing is that this is not simply a matter of hypothetical conjecture. We have clear, demonstrated results from people on the front lines who have proven that this approach works. Part of that comes from the work of people like Admiral Thad Allen, who played such an important role in coordinating the Federal effort in New Orleans. Part of it has to do with lessons taught on the morning of September 11th just across the river here in Arlington County, where Federal, State, and local officials worked together in a remarkably seamless way. It is almost as if, Mr. Chairman, they had read and could have written your report on Hurricane Katrina because they already have demonstrated the lessons of what it is that works.

So, in short, Mr. Chairman, we know what it is that works, and we know that it can be done. We know that what it requires most is strong and effective leadership. A lot of people sometimes say that it is just a matter of rocket science, or it is not rocket science. Well, in a sense it is rocket science because if you look at the ways in which people, in fact, launch rockets, they get people from the different disciplines together in the same room, they work together, they collaborate, they share information and work together under a single command to decide what has to be done, how it has to be done, and make sure that those effective disciplines come together in the way to make the right decision.

In a sense it is rocket science, and in a sense the lessons of rocket science are the same lessons that we learned on the morning of September 11th at the Pentagon. Effective, coordinated response on the part of Federal, State, and local officials is something that we know how to do. What we need to learn how to do is to figure out how to do it more often, more predictably, and more regularly.

Thank you very much, Mr. Chairman.

[The prepared statement of Dr. Kettl follows:]

Testimony

Committee on Government Reform
U.S. House of Representatives

March 30, 2006

Information Sharing
and
Institutional Awareness

Donald F. Kettl

Director
Fels Institute of Government
University of Pennsylvania
3814 Walnut Street
Philadelphia, PA 19104
215.746.4600
dkettl@sas.upenn.edu

With this hearing, the Committee has identified one of the most difficult yet most critical issues that have surfaced since Hurricane Katrina's destructive path through the Gulf Coast: how the large collection of organizations—public, private, and nonprofit; federal, state, and local—can effectively share information to improve the nation's response to predictable natural disasters.

We must learn the critical lessons. In part, unfortunately, future Katrina-scale disasters are a certainty, and we owe it to the nation's citizens to improve our ability to respond. In part, terrorist events would pose precisely the same kinds of problems, and we cannot allow an inadequate response to undermine national security. It is a great honor to appear before you to explore these issues.

A Failure of Initiative—A Failure of Coordination

The House Select Committee pointed squarely to a failure of initiative in responding to Katrina. And, as Committee's report argues, that failure is a failure of coordination. Indeed, government's response to Katrina ranks as one of the worst failures of government administration in the nation's history.

Two things compound that tragedy. First, the nation invested the four years following the September 11 terrorist attacks to ensure that the nation would be ready the next time. It was not.

Second, everyone involved in the response—from top federal officials to local officials on the front lines, from leading nonprofit organizations to private companies—tried their very best. Despite that, citizens needlessly suffered, and some unquestionably died because their government did not serve them well.

Learning these four lessons would build a far stronger, much more nimble system of emergency response.

Lesson 1. Structure: Leadership matters more

The first lesson is that some bureaucratic structures are clearly better than others. But we need to sidestep the inevitable temptation to tinker with the structure and focus instead on getting the fundamental parts to work together better.

The central question is whether the lead government agency, the Federal Emergency Management Agency (FEMA), ought to be moved out of the Department of Homeland Security and be remade into an independent super-agency charged with emergency preparedness and response. If we rolled back the clock, we might well not have put FEMA into DHS to begin with. There is substantial evidence that the move disrupted FEMA's organization and led to the departure of a large number of its most skilled employees.

However, the last thing FEMA needs now is yet another fundamental disruption to its organization and operations. Over the decades, it has bounced like a ping pong ball around the federal bureaucracy. Deciding now to make FEMA independent of DHS would solve some problems. It would just as surely create new ones. And it certainly would stir up more turmoil just as the agency is trying to regain its feet.

There is no reason to believe that a fundamental restructuring of FEMA would solve its coordination problems. A different structure would bring new advantages—and new disadvantages. It would also perpetuate the myth that the complex and wildly varying nature of problems FEMA faces can be solved through structural changes. What FEMA most needs is strong leadership devoted to collaboration, and political support from the highest levels of government—in the White House and in Congress—for this mission.

Consider the boundaries separating FEMA's regions and the path of Hurricane Katrina (see Figure 1). Katrina showed an uncanny instinct for finding the cracks between the FEMA's regional boundaries. If FEMA now restructured its boundaries to prevent a recurrence of the confusion that surrounded Katrina, the next storm might well outwit the structural designers yet again. Biological hazards might well pose very different problems than natural disasters, and terrorist attacks could pose yet more confounding problems.

Structure unquestionably matters. Some structures are better than others. But responding to every problem with a new structure is certain only to destabilize the organization's operations and undermine its ability to respond to new crises. At this point, moving FEMA yet again would divert attention from the more important imperative of building a new coordinating strategy.

There is nothing inherent in FEMA's current structure that prevents its officials from working closely with other officials, in other agencies and at other levels of government. What FEMA most needs now is not another shuffle in the deck of the government's organization charts.

More generally, information sharing and institutional awareness requires strong and effective leadership. Different structures can make the job easier. But people matter most.

- ***We need to strengthen the leadership of our key governmental organizations—and to focus our leaders on building an effective network of action—before we engage in further tinkering with organizational structures.***

Lesson 2. Culture: Build an all-hazard system to bridge organizational boundaries

That closely connects with the second lesson. The central task of leadership is to create an organizational culture that supports the mission.

The investigations into the federal government's response to Katrina show that our heightened concern about terrorism has undermined our ability to deal with other events. In moving FEMA to the Department of Homeland Security, weakening its preparedness function, and failing to build a strong intergovernmental component into its operations, we have made it far harder for FEMA to do its job: not only in responding to natural disasters, but in dealing with terrorist acts as well.

At the core of FEMA's struggle to deal with Katrina was the change in culture that preceded the storm: a focus on terrorism, to the exclusion of other hazards; and an emphasis on response, to the exclusion of remediation and other strategies of reducing risks in advance of events. The narrow focus produced a tunnel vision that dramatically reduced FEMA's capacity to respond to natural disasters like Katrina

The enormous complications of merging 22 different federal agencies into DHS have understandably preoccupied top leaders. They have devoted an enormous amount of energy simply to trying to synchronize the operations of its disparate agencies. That process, however, has created an inward-looking culture with an extremely narrow tunnel vision, devoted to controlling activities dealing with homeland security. Its mission demands a flexible, outward-facing culture devoted to building partnerships with the vast range of organizations—public, private, and nonprofit; federal, state, and local—whose operations, put together, define how well the nation's response works.

This by no means suggests that we should reduce our attention to terrorism. Rather, it underlines the need to pay attention to these important principles, taught by first responders:

- *All terrorist events begin as local events.* On the morning of September 11, New York City's firefighters knew at first only that they were responding to a very large fire in the World Trade Center, probably caused by a plane crash of some kind. The October anthrax attacks began with a mysterious death of a photo editor in Florida. Wisconsin's monkeypox outbreak began with a patient who came to the doctor with lesions that looked like smallpox. Unlike some natural disasters, which appear days in advance on weather maps, terrorist events occur suddenly as an emergency in some community.

- *All natural disasters begin as local events.* The first indication of the failure of the New Orleans levees was when National Guard troops looked outside their door to find the water suddenly rising. Tornadoes, floods, and earthquakes begin as neighborhood-based events. The searing post-Katrina videos as well as the sad tales that accompany other natural disasters, remind us that natural disasters likewise focus their effects on some community.

- *The effectiveness of the response to such events typically does not depend on what caused them.* Many of the canine search and rescue teams that worked in the horrible conditions of Ground Zero in New York joined the search in New Orleans. For the dogs searching for victims, it did not matter whether terrorists or hurricanes had toppled the buildings. For victims trapped on rooftops in New Orleans, it did not matter whether terrorists or flooding had chased them there. When disasters occur, people need help. Some terrorist events, like dirty bombs, raise special first-response issues. In general, however, terrorist events and natural disasters create similar issues for government's response.

An organization's culture defines which issues are seen as problems, which problems become most important, and how the organization responds to the important problems. In recent years, FEMA has focused more on terrorism. That focus transformed its culture and reduced its capacity to deal with natural disasters. Given the agency's performance in the aftermath of Katrina, there is no reason to believe that its response to a terrorist event would have been any better. After all, the agency had several days' notice of Katrina's arrival; terrorists would certainly not provide advance warning.

We need to strengthen our capacity to respond, quickly and effectively, to disasters, whatever their cause. Experts call this an "all-hazard" approach. We need to refocus FEMA's culture, in particular, on all all-hazard approach. Enhancing the capacity to respond to hurricane victims surely does not diminish the capacity to respond to terrorist attacks. Indeed, it would only strengthen it.

The "all-hazard" approach is one that is far more than a homeland security strategy. It is one that challenges agency managers to step back and ask what purposes their organizations seek to achieve. It demands that these purposes, not the narrow constructs of organizational boundaries, shape their behavior and the work of their employees. It is a way of shaping and driving the organizational mission.

- *We need to focus the energy of our top governmental leaders on creating an outwardly looking culture that supports their mission. Internally looking structures, rules, and organizational silos will only prevent our public organizations from getting the job done.*

Lesson 3. Technology: Enhance interoperability

The third lesson is that our fragmented communication systems have made it difficult, sometimes impossible, for first responders to work together in emergencies.

On September 11 and following Katrina's assault on the Gulf, we learned that the central problem in government's response was coordination. After the hurricane hit, coordination problems cascaded. The city of New Orleans had difficulty in cooperating with state officials. Louisiana state officials complained that FEMA was unresponsive. FEMA officials said that they had difficulty in getting the attention of top White House officials. Managers in other federal agencies said offers of help went unanswered or that FEMA made it difficult to get help to where it was needed.

According to the House Select Committee that investigated the response to Katrina,

> Many of the problems we have identified can be categorized as "information gaps"—or at least problems with information-related implications, or failures to act decisively because information was sketchy at best. Better information would have been an optimal weapon against Katrina. (p. 1)

These information gaps led to unfortunate problems of coordination that severely undermined the nation's response to the tragedy.

"Coordination," of course, is the universal goal for all complex operations. "Failure of coordination" is the universal diagnosis for most failures. But the nation has now been taught the same searing lesson twice: the first casualty of many catastrophic events is the regular communication system. And as difficult as coordination is under normal circumstances, it is vastly more difficult—and sometimes impossible—when the communication system fails.

Sometimes the failure is the collapse of the technology. When Katrina hit, it took down radio, telephone, and cellular towers. Many New Orleans police officers were on the streets without any way of communicating with their colleagues or with headquarters.

Sometimes the failure is the technological—the inability of officials in different organizations to talk to each other, often because their radios operate on different frequencies. That problem plagued Louisiana state police officers, who were unable to talk with their colleagues in local law enforcement. It plagued members of the National Guard, who arrived in Louisiana to help but discovered that they could not talk to each other. The National Guard commander finally solved that problem by buying new radios. The same problem afflicted first responders in Mississippi as well.

We will keep paying the painful price of failing to learn the fundamental lesson of September 11: we need robust, interoperable communications systems. A disaster, natural or manmade, will strain our resources under the best of circumstances. But when we do not do what we can to make our basic communications technologies link better, our citizens will unnecessarily suffer.

- *We need to invest our scarce public dollars on plugging the foundation of our national response system: the technological cracks that prevent our first responders from coordinating with each other when disaster strikes.*

Lesson 4. Bureaucracy: Focus on results

The fourth lesson is that bureaucratic procedures, designed to get the public's work done, too often get in the way.

The reports by the House Select Committee and the White House make this point repeatedly. Journalists' stories are legion. Louisiana Governor Kathleen Blanco, in her first phone call to the president, asked for "all federal firepower." She continued, "I meant everything. Just send it. Give me planes, give me boats." New Orleans Mayor Ray Nagin sent out his own plea: "I need everything." He criticized federal officials and said, "They're thinking small, man. And this is a major, major, major deal. And I can't emphasize it enough, man. This is crazy." When top federal officials told them help was on the way, Nagin countered, "They're not here." Frustrated, he added, "Now get off your asses and do something, and let's fix the biggest goddamn crisis in the history of this country." Federal officials said that they were awaiting clear requests, submitted in the proper form, from state and local officials. The difficulty of negotiating these rules meant that help was painfully slow in coming.

FEMA's core problem lies in its inability to secure effective coordination among all of those who help is needed to build an effective response. An organization with a fixed hierarchy and a fixed pattern of response will always be overwhelmed by events that do not match its structure. Since the array of homeland security events—natural and manmade—are unpredictable by their very nature, that approach dooms an organization with such a strategy to failure. That is precisely what happened in Katrina, and it is surely what will recur if we do not develop a better system.

The steps to building an effective system of coordination should not be seen as a problem of structure, which requires a structural solution. Rather, it needs to be seen as an issue of partnership, which requires leadership. The top government official needs to act as the conductor of a well-tuned orchestra, not as the commander of a hierarchy. In Katrina, there was an unseemly fight for the baton.

Rules matter. The violation of federal contracting rules in the hectic days and weeks after Katrina hit led to millions of dollars of questionable expenditures, according to the Government Accountability Office. But rules guide agencies in pursuit of their missions. Results matter most.

How can we create a nimble, effective, robust, results-driven strategy? We need strong leadership.

Such leadership, in turn, requires:
- Establishing FEMA as a reservoir of expertise, for both remediation and response.
- Creating within FEMA the locus of strong command. That command should focus bringing together the needed capacity wherever it can be found—not on insisting on giving orders through a hierarchy.
- Fashioning effective partnerships among the vast array of federal agencies whose expertise and capacity might be needed in a crisis. Not all agencies will be needed in every crisis, and which agencies will be needed when is impossible to predict.
- Building an effective intergovernmental link between FEMA and state/local governments.

This requires coordination that is both vertical (from local and state governments to the federal government) and horizontal (across the range of federal agencies with the ability to contribute to government's response). Such a system must, by necessity be flexible and lithe. It must be based on a networked, not a hierarchical approach to governance. It requires strong leadership to secure coordination. A "center-edge" approach (see Figure 2) provides a model. Consider how it would apply to FEMA and responses to homeland security events.

- *FEMA at the center.* At the center should be FEMA. Its job would be to set policy goals; steer the system to achieve these goals; and measure results. It would provide money to partners in the network, including state and local governments, to reduce risks in advance and to enhance their ability to respond when needed. It would collect information about what works best.

In short, FEMA needs to be the conductor of a well-tuned orchestra, equipped to play the right notes depending on the score—depending on the events it must confront. That requires strong and effective leadership—leadership tirelessly devoted to building effective partnerships.

- *Federal agencies supporting the middle.* In the middle should be other federal agencies. Many agencies have the capacity to contribute to the federal government's response. The Department of Transportation can supply logistical help. Housing and Urban Development can assist with housing. Defense can provide emergency relief supplies, such as food and water, as well as helicopters and heavy equipment. Its forces, including the National Guard and federal troops, can provide needed manpower. In disease-based and bioterror events, the Centers for Disease Control and the National Institutes of Health, among many other agencies, could play an important role.

In short, many federal agencies are potential contributors to a homeland security effort. Which agencies need to get involved depend on the nature of the event. Since events are hard to predict in advance, FEMA needs to be flexible, ready to bring in the assistance it needs, depending on the problem. It needs to be able to do so quickly, reliably, efficiently and responsively.

In each federal agency with an important—or potential—homeland security role, FEMA ought to identify a senior liaison official. This liaison ought to be prepared to deliver that agency's capacity when needed. These relationships ought to be tested and practiced, in advance, through a wide range of all-hazard exercises. The federal government needs to be prepared to respond with what is needed, when it is needed. The problem ought to define the strategy.

- *State and local governments at the edge.* Subnational governments work at the front lines. The first response system will only be as good as their response. FEMA has a central responsibility in ensuring that they are prepared to respond effectively. FEMA also has a central responsibility for bridging the gap between levels of government and between governments at the same level. For example, major communication problems have plagued every major homeland security response in recent years. FEMA has an obligation to help resolve those problems.

To ensure the system's ability at the edge to meet the widest possible array of homeland security problems, FEMA should enhance the role of its regional offices to secure a coordinated response. To do so, FEMA's regional offices should embed senior FEMA staffers in each of the 50 state homeland security offices, and it should work with them to build a coordinated all-hazard strategy.

Finally, FEMA should once again make remediation a major part of its mission, and to make grants to state and local governments a major part of its remediation strategy. In the past, critics have charged that homeland security grants were little more than patronage. In an era of high risk and tight budgets, that is unacceptable. Congress can avoid that problem by focusing the grant system on the areas and issues of highest risk, and by making the grants conditional on achieving high performance.

In short,

- *We need to create a system that rewards results. We do not need a system that obsesses over procedures and jurisdictions. When people are in trouble, they rightly expect their government to help—and they rightly are impatient with excuses built on poor leadership or misplaced devotion to rules.*

Putting it together

This is an imposing agenda. Can it work?

There is, in fact, ample evidence already in place that such a strategy would work effectively. Under the leadership of Admiral Thad Allen in New Orleans, the U.S. Coast Guard has already demonstrated the value of such an approach. The lesson is that the situational awareness that comes from front-line experience provides a guide for solving these problems.

The response of the first responders at the Pentagon on the morning of September 11 teaches the same lesson. Arlington County fire, police, and emergency officials led the response, but they had effective support from the FBI and other federal officials. Mutual-aid agreements with surrounding jurisdictions brought much-needed reinforcements, and Virginia officials worked well with officials from Maryland. Unlike the chaotic and troubled intergovernmental work in the Gulf, this team worked together well because they had developed carefully coordinated plans and they practiced them, together and repeatedly. In fact, many of the commanders at the scene of the Pentagon that Tuesday morning had just they had completed a joint exercise the previous Sunday. They had not anticipated a terrorist attack on the Pentagon by means of a hijacked airplane, but because they had developed and practices working together, they responded predictably to the unpredicted attack.

From these experiences come proven lessons:

- We are facing complex issues that no organization, no matter how it is structured, can fully own or control. A coordinated multi-organizational response is essential.

- Such a response depends on creating action plans in advance—and practicing them repeatedly.

- This practice not only makes interorganizational cooperation second-hand. It also creates personal relationships of trust among the top officials.

- This cooperation depends heavily on developing and sharing information among the principals. A resort to bureaucratic rules and structures hamstrings the response.

- That information sharing needs to be supported with up-to-date, interoperable communication systems. The technology needs to be an important ally, not one more barrier to overcome.

- Information supplements hierarchies. It doesn't replace it. Traditional organizations need to continue to develop strong, effective competencies. They need to come together in flexible, mutually supportive networks, as the mission requires.

- The mission drives the partnership. It defines who needs to play which roles, to meet a wide range of problems, many—perhaps most—of which are unanticipated.

- The more that organizations need to coordinate on complex problems, the more we need highly skilled individual leaders who see such action as job one.

Thanks to the hard work of front-line managers, we know how to do these things. We have evidence that it works. We have struggled with the painful consequences of failing to learn the repeated lessons of past events. We know that we need a system built on these principles:

- *Operational awareness.* Institutions, especially governmental ones, focus on a keen awareness of the front-line operational problems.

- *Coordination.* Government agencies work aggressively to develop an integrated, coordinated response to these problems.

- *Shared information.* That response builds that response on shared information.

- *Focus on results.* Governmental leaders focus on results, not procedures, as the primary standard of accountability.

Figure 1
FEMA's Regional Office Boundaries
and the Path of Hurricane Katrina

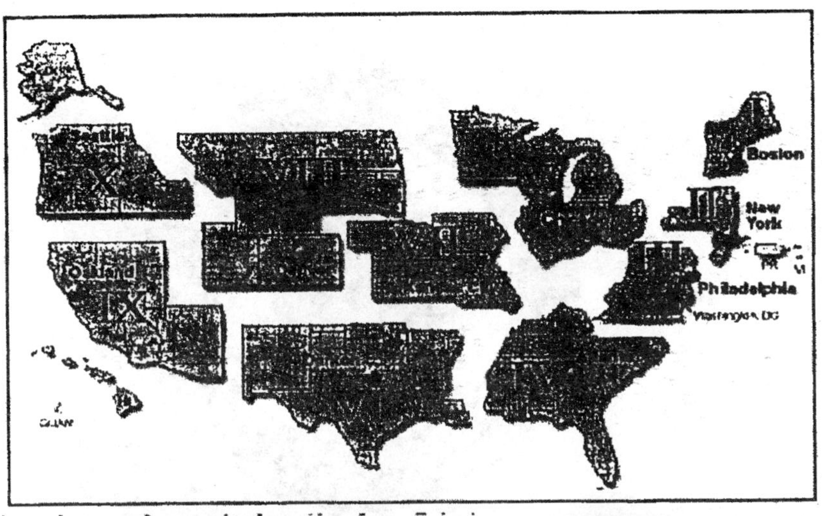

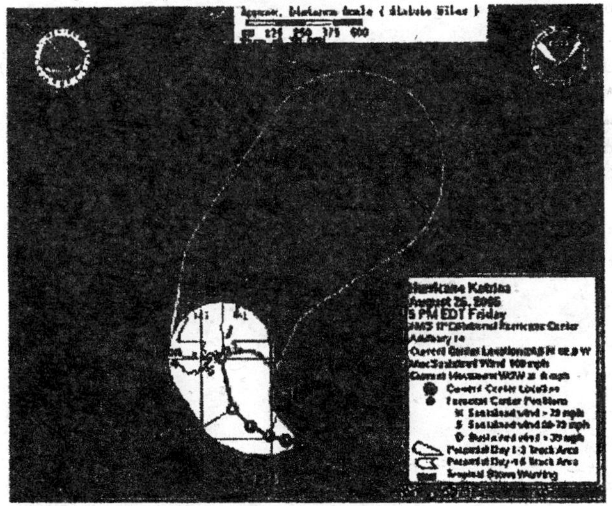

Figure 2
The Center-Edge Approach

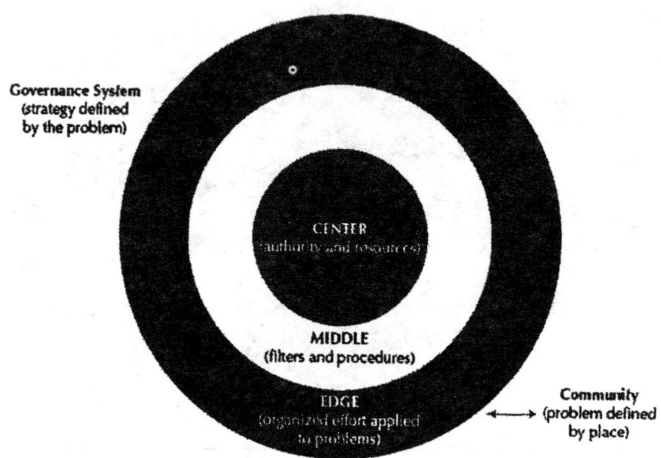

Source: Donald F. Kettl, *The Next Government of the United States: Challenges for Performance in the 21st Century* (Washington: IBM Center for the Business of Government, 2005). See http://www.businessofgovernment.org/main/publications/grant_reports/details/index.asp?GID=235

Chairman TOM DAVIS. Thank you.
Dr. Jackson.

STATEMENT OF BRIAN A. JACKSON

Dr. JACKSON. Thank you. Mr. Chairman, thanks for inviting me to participate in today's hearing. I should begin by saying that my remarks are principally based on our published study entitled, "Protecting Emergency Responders: Safety Management in Disaster and Terrorism Response," which was a joint research effort between the RAND Corp. and NIOSH, the National Institute for Occupational Safety and Health.

The focus of our study was on safety management, which is, of course, a subset of overall disaster management. Many of the recommendations focused on improving safety management are focused on information sharing and are, therefore, very relevant, looking at sort of a specific case within management of an overall disaster. In our study, we looked at four disasters: the two September 11th responses, which have been mentioned previously; Hurricane Andrew in 1992; and the Northridge earthquake, to, again, sort of build on we have been learning these lessons over a long period.

Our work was done in close collaboration with the emergency responder community, including folks who were involved in managing those response operations, and our recommendations were also vetted by other emergency responders, so this is really something that is coming from the responder community.

I really want to focus in on three major lessons to sort of pull out some of the elements from my written testimony.

First, disaster response operations have different levels of information sharing requirements. We have been talking about this as sort of, you know, one topic, but to manage responder safety, for example, the incident commander at the scene needs strategic-level information: what injuries are happening to the responders and what things they can take—changes in the way that the response is done—to keep them safe.

At the tactical level for individual responders, the information sharing requirement is very different. Getting information about what safety actions they need to take to protect themselves. Again, going back to the September 11th response, the question about which respirator to wear when is a very important and operational issue when you are dealing with a large-scale event.

This suggests that there is a requirements generation process that is needed in this to ensure that the information that individual responders, whatever level of safety management they are, gets there when they need it. And also differences that exist across the country, even looking at the four cases that we examined in areas with capable response organizations, imposing a one-size-fits-all sort of solution from the top down, there are risks associated with doing that because of the differences in the way the response organizations structure themselves and manage themselves. Furthermore, sort of the answer of getting all information to everyone at all times, to sort of echo one of the points that was made earlier, is also problematic because if you have to sift the critical information that you need out of a very large background of useful but perhaps not immediately useful information, more sharing may actually result in the information needs of the responders not being met.

Second, the goal is not just getting information there. It is having responders be able to use it when they get there. So the other part of the equation about making sure that the way information is presented to different response organizations at these multi-agency responses is important. The example from the safety case, telling a responder that a certain contaminant is at 20 parts per million in the air may be entirely irrelevant if you do not know whether that is a hazard, or if it is a hazard, what you should do as a response to it.

And then, last, again echoing a point made by other witnesses, although technology clearly has a role to play here and failures in technology can result in bad information sharing, information sharing is really driven in large part by people. In a disaster, managers need to know what organizations to reach out to. If they don't have existing relationships with those organizations, the time-critical point after a disaster is not the time they will be looking for the relationships to build. They have to trust the information that they get back so they can actually act on it and use it in what is generally a life safety situation. And so as a result, having representatives meeting each other for the first time in a disaster working operation is not a good recipe for success.

So as a result, our core recommendation in our report was the need for individuals to play this role of human bridges. We were looking at safety so we focused on individuals we called disaster safety managers. Again, recognizing differences between areas, we did not see this as something that was coming down from the Federal Government, but, again, bringing you back to some of these sort of human network recommendations you heard earlier, safety managers have to be local enough that they have these relationships with the organizations that will be cooperating if a disaster happens in their area, but also have the knowledge to know where and how to reach up to the Federal or other national level organizations that will either be coming to join or support an operation. To us, that suggests that a model sort of designating individuals drawn from either Federal, State, or local organizations where part of their job was to build and maintain those connections.

So, in closing, I would like to thank you again for the opportunity to address the committee today, and I look forward to answering any questions.

[The prepared statement of Dr. Jackson follows:]

TESTIMONY

Information Sharing and Emergency Responder Safety Management

BRIAN A. JACKSON

CT-258

March 2006

Testimony presented before the House Government Reform Committee on March 30, 2006

Published 2006 by the RAND Corporation
1776 Main Street, P.O. Box 2138, Santa Monica, CA 90407-2138
1200 South Hayes Street, Arlington, VA 22202-5050
201 North Craig Street, Suite 202, Pittsburgh, PA 15213-1516
RAND URL: http://www.rand.org/
To order RAND documents or to obtain additional information, contact
Distribution Services: Telephone: (310) 451-7002;
Fax: (310) 451-6915; Email: order@rand.org

Brian A. Jackson[1]
The RAND Corporation

Information Sharing and Emergency Responder Safety Management

**Before the Government Reform Committee
United States House of Representatives**

March 30, 2006

Mr. Chairman and distinguished Members of the Committee: Thank you for inviting me to participate in today's hearing on this important subject. I should begin by saying that my remarks are principally based on the 2004 published study entitled *Protecting Emergency Responders, Volume 3: Safety Management in Disaster and Terrorism Response*.[2] The study was a joint effort of the RAND Corporation and the National Institute for Occupational Safety and Health. I am therefore drawing both on my own work and that of my co-authors, John Baker, Susan Ridgely, and James Bartis of RAND and Herbert Linn of NIOSH, although the specific content of my remarks today is my responsibility alone.

The focus of the study was on safety management – the processes and capabilities needed to keep responders safe at disaster response operations. Our study examined safety management in both major terrorist attacks and natural disasters. Specifically, we studied the responses to the September 11 attacks, Hurricane Andrew in 1992, and the Northridge Earthquake in 1994. Our work was done in close collaboration with the responder community, including individuals involved in managing the disaster responses we examined.

---

[1] The opinions and conclusions expressed in this testimony are the author's alone and should not be interpreted as representing those of RAND or any of the sponsors of its research. This product is part of the RAND Corporation testimony series. RAND testimonies record testimony presented by RAND associates to federal, state, or local legislative committees; government-appointed commissions and panels; and private review and oversight bodies. The RAND Corporation is a nonprofit research organization providing objective analysis and effective solutions that address the challenges facing the public and private sectors around the world. RAND's publications do not necessarily reflect the opinions of its research clients and sponsors.

[2] *Protecting Emergency Responders, Volume 3: Safety Management in Disaster and Terrorism Response*, B.A. Jackson, J.C. Baker, M.S. Ridgely, J.T. Bartis, and H.I. Linn, RAND Science and Technology and National Institute for Occupational Safety and Health, MG-170-NIOSH, June 2004, available at: http://www.rand.org/pubs/monographs/MG170/.

Other elements draw on other parts of RAND's research related to emergency and disaster response included in:

Protecting Emergency Responders, Volume 2: Community Views of Safety and Health Risks and Personal Protection Needs, T. LaTourrette, D.J. Peterson, J.T. Bartis, B.A. Jackson, and A. Houser, RAND Science and Technology Policy Institute, MR-1646-NIOSH, August 2003, available at: http://www.rand.org/pubs/monograph_reports/MR1646/, and

Protecting Emergency Responders: Lessons Learned from Terrorist Attacks, B.A. Jackson, D.J. Peterson, J.T. Bartis, T. LaTourrette, I. Brahmakulam, A. Houser, and J. Sollinger, RAND Science and Technology Policy Institute, CF-176-OSTP/NIOSH, March 2002, available at: http://www.rand.org/pubs/conf_proceedings/CF176/.

Safety management is a subset of overall disaster management, so there are strong parallels between what is needed to effectively protect responders and what is needed for an effective response. Many of the lessons from our work relate to information sharing among responding organizations.

Let me begin by stating the central messages from our study that are most relevant to the Committee's interests in this hearing. Our work identified a range of strategies to improve information sharing at disaster response operations. While interoperable technologies at the disaster scene and planning for information sharing are essential, success is driven in large part by the people involved in managing response operations – the "human bridges" among responding organizations – and, as our case studies showed, the necessary bridges are not always in place when disasters occur. Furthermore, the requirements for information sharing must be addressed during preparedness efforts. Building the needed relationships between individuals and organizations so that information can flow is difficult or impossible in the charged and high-pressure atmosphere of an ongoing disaster response. For sharing to occur effectively, the elements must be in place before a disaster occurs.

The Nature of Disasters Drive the Need for Information Sharing

The scale and demands of disasters require the participation of many different response organizations and specialties. Organizations involved in response operations span many professional disciplines – such as emergency management, fire service, law enforcement, emergency medical services, and responders from other government organizations at the local, state, and federal levels. They also frequently include organizations outside of government and from the private sector. For example, published estimates put the number of organizations involved in the 9/11 response at the Pentagon at over 100 and the number involved at the World Trade Center site at over 400.[3] Managing a disaster response effectively requires bringing together the activities of many disparate organizations into a unified effort, and doing so requires the ability to share information among them.

Information sharing is similarly critical for protecting the safety of responders. After a major hurricane, earthquake, or a large-scale terrorist attack, responders face a wide variety of hazards, which can vary from disaster to disaster. Assessing specific hazards and deploying protective measures may require technical specialists that are not present in all responding organizations.

---

[3] See *Protecting Emergency Responders, Volume 3: Safety Management in Disaster and Terrorism Response*, p. 32 and references therein.

For example, the airborne hazards present at the 9/11 response scenes from the pulverized building materials included a range of hazards that required specialized capabilities to analyze and assess. The uniqueness of post-disaster environments means that many responders may be facing unfamiliar hazards – or at least hazards that are rare in their day-to-day operations – and therefore lack sufficient knowledge to guide their deployment of protective equipment. Information will therefore need to be shared among involved organizations to ensure responders take appropriate protective measures during the response. With respect to responder safety, we refer to this as the need for integrated safety management – where the protective capabilities of all organizations present can be brought together to benefit all the responders involved.

Information Sharing Needs for Protecting Emergency Responders

Effective decisionmaking in disasters relies on having access to the right information at the right time. Though managing overall disaster response has a broader set of information requirements, safety management requires key pieces of information to guide risk management and decisionmaking.[4] Information requirements include:

- *Information about the hazard environment* – Incident commanders and safety managers need to know the hazards responders face, what they mean from a safety perspective, and what can be done about them. If hazard assessment requires specialized technical capabilities, the results of that assessment will need to be shared among responding organizations. When shared, the information needs to be presented in ways that response organizations can act upon it for decisionmaking.

- *Information on the responder workforce* – Incident commanders and safety managers that are making decisions at a disaster scene need to know who is involved in the response and what they are doing. This information is needed to make risk management decisions and, if safety issues arise, locate affected responders so their needs can be addressed. For example, in Hurricane Andrew the large area affected by the disaster and the involvement of many convergent volunteers in the response meant that incident managers had little information on the responders at the scene and their activities. In the absence of central accountability systems for responders, these data are only maintained within individual responder organizations and must be shared to enable coordination of activities at a response.

---

[4] Similar types of hazard information are also needed to enable decisionmaking about measures to protect members of the public who may be located in or near the disaster scene.

- *Information on evolving safety issues* – If responders who are involved at a disaster are being injured or exposed to hazardous environments, commanders need to know about it. Getting such information rapidly is critical to enable deployment of protective measures. In a number of the cases we examined, including operations at the World Trade Center after the September 11 attacks and Hurricane Andrew, response commanders indicated that data on the injuries suffered by responders at the scene were not collected or were not collected in a way where they could be effectively used to address the safety problems that were causing the injuries. If this information was collected at all, it was generally collected within individual organizations and therefore unavailable to the managers of the overall response. Building a common picture of the hazard environment so action can be taken to protect all responders at a disaster therefore requires sharing such information.

- *Information about safety equipment* – Addressing many safety issues requires the use of specific types of protective equipment. Managing safety effectively therefore requires information on what equipment is available and how to match that equipment to the hazard environment at the disaster. The classic example of this issue is the problems of interoperability of safety equipment – whether replacement cartridges for respirators or batteries for other equipment – that affected responders to the September 11 attacks at both the World Trade Center and the Pentagon.[5] In the wake of a disaster, large amounts of equipment and other resources are frequently sent to the area by many organizations acting independently. Effective information management and sharing about available equipment is critical to match the right resource to a safety need. Without it, responders may not be able to find what they need among the stocks of equipment at an incident scene.

However, successful safety management – and, by extension, disaster management – is not just a point of "more information sharing to everyone at all times." Responders have different information needs. The overall incident commander or safety manager of a disaster response operation needs strategic level information to make macro-scale safety decisions for the response as a whole. At the tactical level, individual responders need specific information to guide their own protective and operational decisions. Given the demands of a disaster scene, both responders need to be able to find the information they need quickly from the streams of data available to them. Simply having all information flowing to everyone might actually reduce

---

[5] See *Protecting Emergency Responders: Lessons Learned from Terrorist Attacks*, B.A. Jackson, D.J. Peterson, J.T. Bartis, T. LaTourrette, I. Brahmakulam, A. Houser, and J. Sollinger, RAND Science and Technology Policy Institute, CF-176-OSTP/NIOSH, March 2002, available at: http://www.rand.org/pubs/conf_proceedings/CF176/.

the chances that information *needs* will be met if the result of increased sharing is that critical data must be sifted out of a larger background of less immediately relevant information.[6]

Addressing Information Sharing Needs for Safety Management Requires a Combination of Approaches

Information sharing is not just about interoperability and whether specific technologies can talk to one another. Solutions to other shortfalls must focus on how organizations come together and interact in environments where all have their own specific missions while also contributing to a larger, coordinated response effort. Any approach for information sharing must also be able to be implemented rapidly in the wake of a large-scale disaster that could happen with little or no notice, nearly anywhere in the country.

Our study made several specific recommendations to improve information sharing for safety management:[7]

- Specific plans that list the types of information that need to be collected, what organizations should do so, and how the data should be shared.
- Guidelines and standards that indicate what to do with information once it is shared. The fundamental goal is to improve actions that are taken, not just to increase information flowing among organizations.
- Technology tools for information management and collection including databases, accountability systems, logistics systems, and approaches to track injuries or illness among responders.

However, for such policies and tools to actually improve information sharing during disasters, additional actions must be taken *before* disasters occur. Development of plans for information collection and sharing is not enough. These tools must be incorporated into training and preparedness exercises to ensure that, when called upon, they will produce the desired results. Responders we interviewed during our study indicated that safety management is frequently

---

[6] We have made similar arguments regarding the more specific information sharing issue of communications interoperability – the goal is communication of needed information to the right people, when they need it, and in a way that can be acted upon. If the result of increasing interoperability is "everyone talking to everyone," therefore making actual communication more difficult, the effect on operational effectiveness could be negative rather than positive (see, "Communications Interoperability and Emergency Response," in *Forging America's New Normalcy: Securing Our Homeland, Protecting Our Liberty*, Advisory Panel to Assess Domestic Response Capabilities for Terrorism Involving Weapons of Mass Destruction (The Gilmore Commission), December 15, 2003.)

[7] To place the information sharing recommendations of the study in the context of the other safety management recommendations, we have included the Executive Summary of the report as an Attachment to my written testimony.

given insufficient attention in preparedness exercises, which often focus on the operational elements of response, limiting opportunities to test and practice safety-related information sharing among responding agencies.[8]

Other pre-incident measures – even ones not specifically focused on information sharing – can also reduce the volume and types of information that must be shared during a disaster response. Technology standards are a good example. When pieces of protective equipment and supplies from different manufacturers are incompatible with one another – as was seen for a number of types of equipment in the 9/11 response operations – ensuring that responders have the equipment they need becomes more difficult. Additional information must be shared to track what kinds of equipment are in use, where replacement supplies are available, and how to match them to meet individual responders' needs. Since many different organizations may be involved in managing logistics at a large disaster, this increases the type and amount of information that must be shared among them. Standardized equipment – however that standardization is achieved – eliminates these information sharing requirements.

When an incident occurs, information sharing among organizations more broadly requires that individual groups can "plug into" a common structure and the right people from each organization interact effectively. The National Incident Management System (NIMS) was built on the Incident Command System that came out of the responder community. As such, it provides a common framework for organizations that need to work together during disaster operations. In our work, on responder safety management, we found that NIMS provides an appropriate structure for integrating the safety management efforts of the many organizations involved in a response. But what is not yet in place are the detailed plans – such as those mentioned above – for how, within NIMS, safety information will be shared, especially among responders from local, state, federal, and non-governmental organizations.

Furthermore, while incident management systems define positions and common approaches for organizations, effective information sharing depends on the people that occupy those positions. In our study, emergency responders we interviewed emphasized that the characteristics of the incident commanders and organizational leaders of response operations drive the effectiveness of management and information sharing. They need to know what organizations to reach out to for specific information and must trust those organizations sufficiently that they will act on the information that comes back. The relationships needed to enable information flows are built up

---

[8] See *Protecting Emergency Responders, Volume 3: Safety Management in Disaster and Terrorism Response*, pp. 85-6.

through operational experience and activities that involve many response organizations, such as metropolitan or regional exercises or preparedness programs.

As a result, the final core recommendation of our study was the need for individuals that could fulfill similar roles for safety management in disaster operations. We labeled these individuals "disaster safety managers" in the report, and a key part of their role would be to act as bridges between organizations for information sharing. The recommendation that individuals be designated to play that role grew out of our conclusion that there was no substitute for effective "human bridges" among organizations to share information and coordinate action. The capabilities suggested for the disaster safety manager included significant management and coordination experience, general knowledge of hazards across the range of possible disasters and response operations, knowledge on safety resources and their availability, knowledge about safety processes and decisionmaking, and preexisting relationships with the organizations likely to become involved in a response if a disaster occurred in that manager's area of responsibility.

Though specific details regarding the recommendation on disaster safety managers are not necessarily germane to a broader discussion of information sharing, there are a few points that are relevant. First, the fact that major disasters are thankfully rare means that few individuals build up all the experience needed to play these roles effectively, even over a career in emergency response. One cannot expect that every response organization will have individuals that can simply step in and do this if a disaster occurs. As a result, we believe there is a need to train individuals to fill these roles. In addition to ensuring that individuals with the right skills are available when a disaster occurs, the training process itself provides a pre-incident information sharing mechanism for relevant lessons-learned to improve future operations. Second, the nature of information sharing needs in disaster operation mean that the individuals playing this role must balance and coordinate in many directions. They must be local enough to have built the connections and relationships with the response organizations in their areas but broad enough to connect with the federal and other national level organizations that will join or support operations. To us, this suggested a model of a small number of individuals – drawn from federal, state, or local response organizations, or some combination thereof – where part of their job function was to build and maintain the connections between organizations to facilitate safety management information sharing in a disaster response.

Conclusions

The effective sharing of information among organizations during a disaster response operation is critical to its success. While interoperable communications and other technologies are one element of effective information sharing, the most critical elements are the "human bridges" that link response organizations. These bridges cannot be built overnight. Development of the necessary human resources and relationships requires significant training and interactions.

It is self-evident that coordinating and drawing on the strengths of all organizations involved will increase the effectiveness of the operational response to a disaster. Similarly, responders are better protected when there is a coordinated safety management effort, rather than relying only the efforts of separate organizations to protect their own members. This is impossible without effective sharing of the necessary information among responding organizations. Improved responder safety therefore represents an added potential payoff to addressing information sharing concerns.

I would like to thank you again for the opportunity to address the committee today, and look forward to answering any questions you might have.

Chairman TOM DAVIS. Lieutenant Lambert.

STATEMENT OF STEVE LAMBERT

Mr. LAMBERT. Good morning, sir. I am Steve Lambert. I am a lieutenant with the Virginia State Police and the agent in charge of the Virginia Fusion Center. Thank you for the opportunity to testify today in this important process, and I look forward to answering any questions you may have at the end of this testimony.

After September 11th, law enforcement agencies were forced to rise and meet the informational demands created by the increased focus on terrorism. The resources needed to provide proactive intelligence operations have increased exponentially. This mere fact has compelled many States and regions to develop Fusion Centers that bring together key critical response elements in a secure, centralized location in order to facilitate the sharing of counterterrorism intelligence information.

Virginia now has such a center with the primary mission of fusing together key counterterrorism resources from local, State, and Federal agencies, as well as private industry, in an effort to prevent the next terror attack. Our second mission, in support of the Virginia Emergency Operations Center, is to centralize information and resources to provide a coordinated and effective response to a terrorist attack or a natural disaster.

It is our contention that having a Fusion Center does alleviate much of the previous resistance to sharing information that has plagued Government response in the past. This business of where to get needed information or just what is available or who can I depend upon for such information can be a terribly confusing process to most any Government or private agency. The bottom line is that Fusion Centers provide a fundamental environment necessary for Federal, State, and local governments to have the proper intelligence and situational awareness to perform their jobs.

Furthermore, and perhaps most importantly—it has been mentioned several times—Fusion Centers are conceptualized to provide the environment of trust between locals to State and State to Federal Government agencies. This issue of trust is absolutely essential. All methods, policies, principles, and techniques are rendered useless if trust is not established between these partners. So essentially the fusion process has created horizontal and vertical bridges for information and intelligence sharing.

To answer the question the committee is particularly interested in—"Are impediments to more effective information sharing primarily technological, or structural, cultural, and bureaucratic in nature?"—the answer from our perspective is that the Fusion Center concept provides a structural solution. It also provides the all important cultural or trust solution. It also provides somewhat a bureaucratic solution and to some extent a technological solution. However, there still exists a foundational and technological hindrance that applies to effective disaster response.

As you know, part of the intelligence process involves identifying gaps in intelligence, and with that, and to my understanding, only a few States have achieved a truly single statewide real-time information and intelligence sharing platform. Although the Fusion Center has taken significant strides toward centralizing this process, there still exists a serious lack of centralized analysis and dissemination function on all criminal intelligence. We all know that good terrorism prevention is good crime prevention and vice versa. However, and like many States, Virginia currently has a statewide information sharing system that suffers from poor participation due to being totally law enforcement centric—excluding all crimes and all hazards—and running on an antiquated architecture. There are simply too many silos. Too much criminal information is being shared by word of mouth and through personal relationships rather than on a single, Web-based, real-time information sharing platform.

The solution to this foundational problem, however, provides tremendous opportunities to revitalize the intelligence process by providing training and including eventually all Virginians in the intelligence process. Taking advise from the 9/11 Report, Virginia has planned to adopt, "a decentralized network model, the concept behind the information revolution, that shares data horizontally too. Agencies would have access to their own data bases but those data bases would be shared across agency lines. In this system, secrets are protected through the design of the network and an information rights management approach that controls access to the data, not the access to the whole network."

Therefore, and in conclusion, how can we avoid the inadequate information sharing and murky situational awareness that characterized the governmental response to Katrina? Establish a Fusion Center or Fusion Centers built on the foundation of a truly integrated, Web-based, statewide information sharing platform that includes all crimes and all hazards.

Thank you very much, sir.

[The prepared statement of Mr. Lambert follows:]

The Honorable Tom Davis
Chairman
U.S. House of Representatives
Committee on Government Reform
Thursday, March 30, 2006
Rayburn House Office Building

Hello, my name is Steve Lambert, I am a Lieutenant with the Virginia State Police and I am the Agent in Charge of the Virginia Fusion Center. Thank you for the opportunity to testify today regarding this important process. I look forward to answering any questions posed by the Members of this Committee at the conclusion of this testimony.

After the September 11, 2001 terrorist attacks, law enforcement agencies were forced to meet the informational demands created by the increased focus on terrorism. As a result, the resources needed to provide proactive intelligence operations have increased exponentially, thus compelling law enforcement agencies to consider the concept of a Fusion Center. This concept envisions bringing key critical response elements together in a secure, centralized location in order to facilitate the sharing of counter-terrorism intelligence information. The Governors Office and the Secure Virginia Panel adopted this concept, thus creating the Virginia Fusion Center (VFC).

Virginia House Bill 1966 required the Governor to establish a multi-agency fusion intelligence center to receive and coordinate terrorist related intelligence. The center shall be operated by the Virginia State Police in cooperation with the Virginia Department of Emergency Management. Virginia House Bill 2032 provided that the Virginia Department of Emergency Management shall be responsible for coordination, receipt, evaluation, and dissemination of emergency services intelligence and shall coordinate intelligence activities with the Virginia State Police. These House Bills laid the foundation for the Virginia Fusion Center and ultimately provided guidance in developing the policies and procedures utilized by the Virginia Fusion Center.

The primary mission of the Virginia Fusion Center is to fuse together key counter-terrorism resources from local, state, and federal agencies as well as private industry in a secure, centralized location, to facilitate information collection, prioritization, classification, analysis, and sharing, in order to better defend the Commonwealth against terrorist threats and/or attack and natural disasters. The secondary mission, in support of the Emergency Operations Center, is to centralize information and resources to provide a comprehensive, coordinated, and effective response in the event of a terrorist attack or natural disaster, in order to neutralize the situation and allow an effective recovery process to be implemented and managed.

One central purpose of a Fusion Center is to remove the resistance to sharing information that has plagued government response in the past, thereby pooling together information from all pertinent intelligence sources to effect a decisive response. In this effort, the Virginia Fusion Center has developed new partnerships and strengthened existing relationships. These new partnerships include private industry and also representatives of local, state, and federal government agencies having a mission critical role in homeland security, such as the health and transportation sectors. The existing relationships that we are strengthening include law enforcement and the military. These partnerships provide the foundation for the Virginia Fusion Center and as they grow, so will the ability to exchange critical information and intelligence to all partners.

The Virginia Fusion Center is comprised of the Information Classification Unit (ICU) and an analytical unit. The ICU receives all incoming information and ensures that it is prioritized, classified, and disseminated in an efficient and timely manner to those personnel and agencies having a mission critical interest in the information. The analytical unit will ensure the proper and timely analysis of the information, thus leading to beneficial intelligence products that can be disseminated to the various partners of the VFC. Currently the analytical unit is comprised of representatives from the Virginia State Police and the Virginia National Guard, but in the near future will have representatives from the Virginia Department of Emergency Management, the Virginia Department of Fire Programs and the Federal Bureau of Investigation. The Virginia Fusion Center also has a representative in the Homeland Security Operations Center to coordinate the exchange of information with the Department of Homeland Security. A Virginia State Police agent and analyst are also assigned to the National Capitol Region Information Center, which is a collaborative effort between the Fairfax Police Department and the Federal Bureau of Investigation.

The Virginia Fusion Center also provides a work area for other federal, state, or local agencies that could have a response or recovery role during a terrorist threat/attack or natural disaster. These agencies have selected representatives that will be able to respond to the VFC as well as provide specific analysis concerning their field of expertise.

Certainly, the fusion process and state established fusion centers are tremendously effective methods of encouraging and enhancing information sharing among regional and even national entities. State Fusion centers can provide the fundamental environment necessary for federal, state and local governments to have the proper intelligence and situational awareness to perform their jobs. Furthermore, Fusion Centers are conceptualized to provide an environment of trust between locals-to-state and state-to-federal government agencies. This issue of trust is absolutely essential. All methods, policies, principles and techniques are rendered useless if trust is not established between these partners. The Virginia Fusion Center has seen this paradigm materialize firsthand and is working very hard to maintain these trusted partner relationships with federal and local agencies.

To answer the question the Committee is particularly interested in, *"Are impediments to more effective information sharing primarily technological, or structural, cultural, and bureaucratic in nature?"* The answer(s) are, from our experience, that the Fusion Center concept provides a structural solution, it provides the all-important cultural or trust solution, it provides a bureaucratic solution and to some extent, it provides a technological solution. However, there still exists a foundational and technological information sharing hindrance for many states and their Fusion Centers that does apply to effective disaster response.

Only a few states have achieved a truly single, statewide, "real-time" information and intelligence-sharing platform. The 2005 National Intelligence Sharing Plan sponsored by the Department of Justice revealed a number of impediments that, from our experience, are certainly true nationwide and also applies to Virginia. Although the Virginia Fusion Center has taken significant strides towards centralizing terrorism intelligence sharing and providing situation awareness to planners, first responders and emergency managers, there still exists a serious lack of centralized analysis and dissemination function on *all criminal intelligence.* We all know that good terrorism prevention is good crime prevention and vice versa. However, and like many states, Virginia currently has a statewide information sharing system that suffers from poor participation due to being totally law enforcement centric, (excluding "all crimes – all hazards") and running on an antiquated architecture tied to encrypted router boxes. There are still too many silos. Too much criminal information sharing is still being done via word of mouth and through personal relationships rather than on a single, web-based, real-time, statewide information-sharing platform.

The solution to this foundational problem facing our fusion process, and other states like us, provides tremendous opportunities to train eventually and potentially all Virginians on the intelligence process and provides avenues to be actively involved in the information/intelligence sharing process. Taking advice from the *911 Report*, Virginia has planned to adopt: "*A decentralized network model, the concept behind the information revolution, that shares data horizontally too. Agencies would have access to their own databases but those databases would be shared across agency lines. In this system, secrets are protected through the design of the network and an "information rights management" approach that controls access to the data, not the access to the whole network.*"

How can we avoid the inadequate information sharing and murky situational awareness that characterized the governmental response to Katrina? Establish a fusion center(s) built on the foundation of a truly integrated, web-based, statewide information-sharing platform that includes all-crimes, all hazards.

Thank you for this opportunity to provide input into this incredibly important process.

Chairman TOM DAVIS. Well, thank you very much.

The votes beat us to it. What I am going to do is take a 20-minute recess, and we will come back and try to move through the questions in short order.

So I will declare a 20-minute recess, and we will be back. Thank you.

[Recess.]

Chairman TOM DAVIS. The committee will reconvene.

Dr. Jackson, let me start with you. Despite the existence of the Hurricane Pam exercise, Katrina showed how even when you predict a disaster, you train for it, it almost always is not sufficient. How do you get people at all levels of Government to get on the same page for preparedness and training?

Dr. JACKSON. Well, in talking to the responders in our research process, the answer that we got from them about that is that it is not a single exercise. It is relationships built over time.

One of the issues about the safety area in particular is that, in contrast to information sharing areas where you can articulate the information that you want to share beforehand, in the safety area it is entirely dependent on the nature of the disaster. So you have to be able to be flexible to reach out through relationships that you perhaps would not have thought would be important beforehand. And so, really, the only answer to that is sort of, you know, repeated interactions between responders during preparedness activities, in exercises. The experience at the Pentagon was cited earlier by one of the panelists. That is an example where that repeated experience over time and the fact that the responders involved had built up those relationships and trust meant that they could adapt flexibly and have the operation go much more smoothly.

Chairman TOM DAVIS. So it's like any teamwork, isn't it? You do your training and your training and your training, and one session does not do enough to create the kind of teamwork.

Dr. JACKSON. Absolutely. You play like you train. And, you know, on these relationships, you know, when—especially, there will always be people who rotate in and out of jobs, you know, within the Federal Government, within the State responder, local responder organizations. There are people who get promoted and move on. And so you need this ongoing process over time, because even if you buildup the relationships today and they are perfect, if, you know, three of those people go on to be promoted and take other jobs, you need to do it again tomorrow.

Chairman TOM DAVIS. So it is practice, practice, practice.

Dr. JACKSON. Right.

Chairman TOM DAVIS. Dr. Kettl, in your opinion, has the Department of Homeland Security sufficiently integrated the local and State emergency management functions to ensure a coordinated emergency response?

Dr. KETTL. Among the many concerns, Mr. Chairman, I have about the Department of Homeland Security, my biggest concern is the lack of integration of State and local issues into the Department of Homeland Security. To be fair to them, they have an enormous challenge in trying to bring 22 different agencies together into a coordinated whole, but the fact is that all homeland security events begin as local events. And the instinct, as unfortunately we saw in Katrina, is not to view State and local responses as critical or integral to their operations. It is perhaps the next generation of responses, but it is a generation that needs to be sped up enormously.

If there is anything that the Department of Homeland Security needs most to do is to devise a far more effective partnership with State and local governments.

Chairman TOM DAVIS. Yes, I agree with you. Just trying to take 170,000, 180,000 employees in 22 agencies, different cultures, different systems, different silos, I think sometimes our expectations are out of whack to expect that to work overnight. And we saw with Katrina that just their own internal communication was not what it ought to be.

Dr. KETTL. I fear that is right, Mr. Chairman. But the point—and this is the source of greatest worry—is that process of trying to integrate all of these complex pieces together has created a kind of top-down approach within Homeland Security, which is understandable. But in the end, Department of Homeland Security operations will only work if they are real from the bottom up and show a sense of operational awareness. And we learned the hard and painful way in the aftermath of Katrina that those instincts, unfortunately, are not there.

Chairman TOM DAVIS. Right, and that was an unforgiving storm.

Mr. Brennan, you testified that we lack a cohesive national framework for emergency response. Have you looked at the National Response Plan, which really never had a chance to be implemented with Katrina because we had——

Mr. BRENNAN. Yes, Mr. Chairman, I have.

Chairman TOM DAVIS. What do you think it needs? To be enhanced? Scrapped? What are your thoughts on it?

Mr. BRENNAN. It is a very bulky document that I think a lot of people do not understand, and it has not really been absorbed within the Federal Government or beyond. I think there are some good ideas and concepts in there, but it also runs afoul of some of the existing statutory responsibilities, authorities, and there are a lot of differences of view about the roles and responsibilities of individual departments and agencies even under that National Response Plan.

Chairman TOM DAVIS. Also, I mean, if you do not train on it, it is such a big plan you are not going to wait for the storm to hit and then read the plan in terms—if you do not train on it—right?—if you do not practice on it, it is not going to do you really any good when the big storm hits, is it?

Mr. BRENNAN. Right. I think it is—as difficult as it was to draft a document like that, it is much more difficult to implement it. It is like a piece of legislation. You know, as difficult as it is to get it through the legislative process, actually operationalizing it is a far cry from passage of that legislation.

Chairman TOM DAVIS. Do you think it has too much flexibility, or do you think it is too prescriptive? Do you have any thoughts on that?

Mr. BRENNAN. It has been a while since I have looked at it, and I think now is the time, after Katrina, to take a really hard scrub at it and see why aspects of it did not work. But I think some of the underlying structures that it really would need in order to be realized still are absent.

Chairman TOM DAVIS. I would just tell you that I know the problem we had with Katrina was that Michael Brown did not believe in the National Response Plan, because in Florida in 2004, an election year, a key State, he was given kind of carte blanche to do what he needed to do. He was talking directly to the White House. The National Response Plan changes all that. He has got to go up through a chain of command, and he was not used to that and did not think he needed to do that. And it seemed like about halfway through, all of a sudden the White House is saying, look, you better go through channels on this. That led to frustration, and the e-mails show that we just kind of crumpled under that.

Mr. BRENNAN. Structure, discipline, and institutionalization of these efforts really is just a prerequisite to actually making things work well in emergency situations.

Chairman TOM DAVIS. Now, you have had experience with DIA and the FBI and other intelligence agencies. What strategies and tactics do you think are the most effective in getting everybody to play ball?

Mr. BRENNAN. Well, there are many different aspects of the ball game. On the information sharing side, in my testimony I talked about the importance of having a common information sharing environment. When I set up the TTIC and the NCTC, we had something called TTIC Online and then NCTC Online that all the different stakeholders would be able to provide information to. So it was a one-stop shopping.

And I think if you take it away from a single department solution or a single functional sort of area, you know, what—it is not a defense issue. It is not an FBI issue. It is not a law enforcement issue. It is not even a single strata issue, as far as Federal, State, or local. You need to have something that is going to bring things together, and there are many different aspects of it: information sharing, communication that we have talked about, command and control. And that is why I really do think a lot of our governance structures and institutions are very much outdated to deal with 21st century problems.

Chairman TOM DAVIS. OK. Thank you very much.

I think one of you mentioned and in the previous panel they mentioned about how too much information could be a dangerous thing. How does too much information hurt you? Just the ability to sort it out and prioritize? I mean, can somebody explain that to me?

Dr. JACKSON. Well, I was one of the people that echoed the earlier panel. I mean, too much information is a problem if what is important gets lost in the flow of it and you can't pick it out.

Chairman TOM DAVIS. Right.

Dr. JACKSON. You know, a lot of our focus in our research was at how to protect individual responders at the lowest level. So, you know, you have a responder who is taking operational action. They have a lot of missions to accomplish at a disaster. They want to know what they need to know, when they need——

Chairman TOM DAVIS. Like you say, charge that hill.

Dr. JACKSON. Yes.

Chairman TOM DAVIS. Without getting into the foreign policy and all that kind of stuff behind it.

Dr. JACKSON. Yes. And if you have to sort of pull out what piece of equipment you should be wearing and what exactly you should be doing from, you know, an entire tome describing everything at the event, you are not—actually, your need, information need, is not being met even though the information has been shared.

Chairman TOM DAVIS. OK. Lieutenant Lambert, let me ask you a couple questions. You noted that the creation of the Fusion Center really breaks down the resistance to information sharing that is ubiquitous in Government. In your experience, what has been the key to successfully pursuing that new approach to information sharing?

Mr. LAMBERT. It seems like we are singing the same chord of trust, having the organizations that are represented, whether it be the FBI, the National Guard, the Department of Emergency Management programs, whoever the first responders are or the information sharers in the same room together in the same building. Actually building on personal relationships I think is probably, at least from my experience, the most important thing we can do.

Chairman TOM DAVIS. You know, Gaebler and Osbourne wrote a book a few years ago called "Reinventing Government," and they have a chapter on mission-driven Government versus regulation-driven Government. And one thing we found in the Katrina investigation is when the military came in, all of a sudden things got down because they were mission-driven. When we saw FEMA and everybody else there trying to go by the book and everything else— and I guess relationships play a role in that. But as we drill on these issues, as you practice and so on, are we doing enough preaching about accomplishing the mission? Or do we preach don't violate the rules and the regulations? Anybody have a thought on that? Dr. Kettl.

Dr. KETTL. Mr. Chairman, I think that is exactly the right point because it both gets to the question of how to deal with the avalanche of information that comes down as well as the question of how you bring different pieces together.

What we know is that operational awareness tends to frame the nature of the problems that have to be solved. If you can get people to agree on what problem has to be solved, it is much easier to bring the pieces together, and it is a lot easier to deal with the process of breaking down the stovepipes if everybody understands what their contribution is to evacuating people off of roofs when they are surrounded by floods, how to get food to people who are hungry, how to deal with avian flu. If the problem drives the solution, it defines the players who need to be involved. It focused them on the nature of the result. And to the degree to which you can get people focusing on that instead of procedures, rules, and structures, coordination is much, much easier. The lesson that people in the first response community over and over and over again is focus on the problem, allow that to drive the nature of the partnerships, and it is a lot easier to then get past the bureaucratic boundaries that so often hamstring action.

Chairman TOM DAVIS. We have a section of the Katrina report where we talk about some of the unsung heroes, and a lot of these people, they were not going by the rules and regulations. We had one doctor who literally broke into Walgreen's to take what drugs were there before they became flooded. He got out of there and walked out with his bag so he could help people who had left home without their prescriptions and the like. We had other folks that were commanding boats that were just hanging around and that would have been flooded out otherwise, basically very, very mission-oriented. Even when you see the action movies, you never saw Steve McQueen or anybody look at the rules and regulations to get it done.

Now, there is a fine line between being mission-oriented and abusing the rules for other purposes and so on. So, you know, we do the oversight on contracts and everything else. We have to recognize that in an emergency situation sometimes the rules need to be relaxed.

I don't know how you preach that, but maybe it is the trust between all the elements that you discussed, the fact that they practiced and drilled together and have relationships which makes a difference and helps you define reasonable boundaries in times of crisis. But that seemed to be a lot of the problem with Katrina. You had the elements working together, but did not trust each other. They knew what—they sort of knew what the mission was. They were told what it was. But at the end of the day, even though we had prepositioned more assets than any other storm in history, it was not near enough. This storm was not just predicted. What happened was predictable, but nobody really got it. I think there was a lot of jockeying around for position and so on, but the storm, which was predicted with absolute—it was absolute in terms of what they predicted, the category, where it would land, but the folks down below really did not get it. And even though they had gone through Hurricane Pam, but you did not have that string of existing relationships that could have made a big difference in this case. This was an unforgiving storm. You make a mistake. It gets exaggerated just because of the size of it, and then the ensuing flooding.

Let me ask Lieutenant Lambert another question. Altering Government agencies' perceptions of information sharing, viewing it as a benefit to everybody, as opposed to giving up turf, if you understand what I am saying, it is the biggest obstacle at the Federal level that we have to overcome. It may be a little easier at the State level to get people working together. You have a strong leader. You have Governor Kaine, let's work for the team. At the Federal level, it is a lot more difficult. You have a lot of entrenched career people that have survived a lot of administrations. Even on Capitol Hill, turf and jurisdiction drive this place to a great extent. A lot of good does not happen because people are nervous about what their jurisdictional battles are going to be in the future of their committees.

What challenges have you faced in this area of trying to get around the perceptions of information sharing and turf battles? Have you had any firsthand experience with that in Virginia?

Mr. LAMBERT. Well, I submit that the same turf battles that the Federal Government experiences also the State government experiences as well. And we have had to take measures to try and develop trust among the locals, State to local. So I can appreciate what they are going through.

I know we went through a time that for some time, just trying to figure out who was organizing Federal intelligence, that we might relate with them rather than dealing with so many different Federal agencies. I think we have—and to DHS' credit here lately, they have really reached out to us, and we have even started a pilot of three more information portals along with the possibility of putting someone in the Fusion Center to, again, strengthen those personal relationships.

But you are absolutely right. It is difficult to overcome all of the bureaucracy.

Chairman TOM DAVIS. It helps to have George Foresman up here, too, in Washington, doesn't it?

Mr. LAMBERT. It doesn't hurt. Yes, sir.

Chairman TOM DAVIS. Dr. Kettl, many have suggested that FEMA be at the center of homeland security events. But FEMA really was not designed to be a first responder or even coordinate the first response. Isn't that right? Can you have FEMA at the center of all operations without enlarging its original scope?

Dr. KETTL. FEMA's role, Mr. Chairman, has changed dramatically over time, and its organizational structure has changed along with it. It is clear that somebody needs to be in a role of playing the central coordinating function. I think of it as kind of a conductor of an orchestra, that you can have a variety of different instruments that appear before you, creating all kinds of different instruments depending on the score that orchestra is trying to play, and the key is having an orchestra conductor skilled enough to be able to play Beethoven one night and Bach the next.

The problem is that FEMA does not see its job as either that orchestra conductor or it does not have the skills for figuring out how to do it. Somebody has to do the job.

Chairman TOM DAVIS. And it should be Federal, right?

Dr. KETTL. It should be Federal, and FEMA is as good a place as any to put it. Now, to do that would require, first, recognizing that is its job; second, getting the political support both from Members of Congress and from senior administration officials to define that, in some cases to provide some additional resources, but then to provide a lot of extra support and leadership essentially to make Lieutenant Lambert's job easier. FEMA's job ought to be to make Lieutenant Lambert's job work better, to try to provide better response in situations like New Orleans.

Chairman TOM DAVIS. In the case of Katrina, Michael Brown was not just the head of FEMA. He was the Federal officer in charge. He was designated—he took it as a demotion, by the way, when it was given to him. And there probably should be that overlap between FEMA and the people being in charge on the ground, but it may be new to FEMA in the sense that they are not necessarily used to this. They were used to coming in 2, 3, 4 days later and doing the mop-up work.

OK. Well, I appreciate that. Is there anything else anybody wants to add?

Mr. BRENNAN. If I could make just one comment?

Chairman TOM DAVIS. Sure.

Mr. BRENNAN. Talking about mission, the challenge is that there are multiple missions that are underway in any type of national disaster or challenge. And it is an unprecedented systems integration challenge that you have law enforcement, you have rescue and recovery, you have security, you have information sharing, you have policy. And my experience has been that there are a lot of disputes about who actually has that statutory authority to exercise command and control over disparate mission elements that are outside of individual departments and agencies that go beyond the Federal area. And that is one of the things that I think is going to continue to be a challenge for, you know, natural disaster response.

Chairman TOM DAVIS. Thank you very much.

I am interested in having Ambassador McNamara, who is the new program manager for the information sharing environment, testify before this committee. I know he is just getting settled into his new position, but is interested in appearing as soon as possible, and given his important role in information sharing across Government and the committee's role in setting the government-wide information policies, we would like him to appear here first when he is able to do so.

I again want to thank this panel and the previous panel. It has been very, very helpful to us. Hurricane season begins officially June 1st, although it begins when it begins. And, you know, who knows whether disasters may strike, and we need to be ready for them. And I hope we have learned the lessons, and I hope this testimony, the administration will take it seriously. I know this committee does.

Thank you very much. The hearing is adjourned.

[Whereupon, at 12:18 p.m., the committee was adjourned.]

○

APEX
TP_CF

9780160766176